Erwin Dee Kord (Ed.)

Nayla Tueni

Erwin Dee Kord (Ed.)

Solv

Nayla Tueni

Ghassan Tueni, Lebanese University, Gebran Tueni, Lebanese Forces

Solv

Imprint

Permission is granted to copy, distribute and/or modify this document under the terms of the GNU Free Documentation License, Version 1.2 or any later version published by the Free Software Foundation; with no Invariant Sections, with the Front-Cover Texts, and with the Back- Cover Texts. A copy of the license is included in the section entitled "GNU Free Documentation License".

All parts of this book are extracted from Wikipedia, the free encyclopedia (www.wikipedia.org).

You can get detailed informations about the authors of this collection of articles at the end of this book. The editors (Ed.) of this book are no authors. They have not modified or extended the original texts.

Pictures published in this book can be under different licences than the GNU Free Documentation License. You can get detailed informations about the authors and licences of pictures at the end of this book.

The content of this book was generated collaboratively by volunteers. Please be advised that nothing found here has necessarily been reviewed by people with the expertise required to provide you with complete, accurate or reliable information. Some information in this book maybe misleading or wrong. The Publisher does not guarantee the validity of the information found here. If you need specific advice (f.e. in fields of medical, legal, financial, or risk management questions) please contact a professional who is licensed or knowledgeable in that area.

Any brand names and product names mentioned in this book are subject to trademark, brand or patent protection and are trademarks or registered trademarks of their respective holders. The use of brand names, product names, common names, trade names, product descriptions etc. even without a particular marking in this works is in no way to be construed to mean that such names may be regarded as unrestricted in respect of trademark and brand protection legislation and could thus be used by anyone.

Cover image: www.ingimage.com
Concerning the licence of the cover image please contact ingimage.

Publisher:
Solv is a trademark of
International Book Market Service Ltd., 17 Rue Meldrum, Beau Bassin, 1713-01 Mauritius
Email: info@bookmarketservice.com
Website: www.bookmarketservice.com

Published in 2011

Printed in: U.S.A., U.K., Germany. This book was not produced in Mauritius.

ISBN: 978-613-8-96310-3

Contents

Nayla_Tueni

Nayla Tueni	
Born	1982 August 31stAchrafieh, Beirut
Residence	Beirut
Nationality	Lebanese
Occupation	Journalist, Politician
Home town	Beirut
Term	(2009-2013)
Religion	Greek Orthodox
Spouse	Malek Maktabi (2009-)
Parents	Gebran Tueni,Mirna Murr
Relatives	Marwan Hamadeh(Great-Uncle) Nadia Tueni(Grand mother) Ghassan Tueni(Grand-father) Michel Murr(Grand-father)

Nayla Tueni Maktabi (Arabic:) (born 1982 August 31, Beirut) is a Lebanese journalist and politician. She won the Greek Orthodox seat for Achrafieh in the 2009 Lebanese Elections for the March 14 block. She and Nadim Gemayel are currently the youngest members of Lebanese Parliament and she is one of the few elected female politicians in the country.[1]

Tueni is a fourth generation journalist. An-Nahar was established by her great-grandfather, Gebran Tueni, in 1933. Her grandfather, Ghassan Tueni, ran the newspaper for decades. She is heiress along with her siblings to the newspaper dynasty and currently a member of the Board and the Deputy General Manager of An-Nahar Newspaper. She is also a board trustee of the *Mentor Arabia*, a non-governmental regional organization that promotes drug prevention and raises awareness on various issues among Arab youth.[2]

Early life

Nayla Tueni was born on August 31, 1982, in Achrafieh. Her family is a prominent Orthodox clan in Lebanon. She pursued her primary education at Collège Louise Wegman and went to high school at Collège Notre-Dame de Nazareth in Achrafieh. She inherited her love and passion for Journalism from her father Gebran Tueni and her grandfather Ghassan Tueni. In 2005 she obtained a Bachelor of Arts in Journalism from the Lebanese University, before earning a Masters degree from the Jean Monet Faculty in Paris. She began her journalistic experience well before graduating from university. In 2003 she was already working as a trainee for An-Nahar and then wrote in the Education and Youth section of the newspaper.[3]

Political views and March 14 alliance

Nayla Tueni is a secular Christian Greek Orthodox. At the time of her political campaign, she stood for belief in Christian preservation, as well as her stance against the Hezbollah, brought her the support of conservative Christians of the Lebanese Forces and Kataeb party. However, she is also an advocate for Secular reforms in the Lebanese sectarian system. In a reportage broadcasted by Al-Jazeera about the power struggle of Lebanon's Christian clans, Tueni stated in Arabic: "The current system means I only represent my sect. I would rather represent my country and not my sect and hope that one day Lebanese politics will not be based on sect".[4]

Among the 48 points she has listed in her political platform are her opposition to the foreign occupation of Lebanon by Israel or Syria, the support of a free democratic Lebanon and political pluralism. She also wishes to see Lebanon remain aside from regional tensions and conflicts, which have turned it into a battlefield for the wars others fight against each other. Tueni also wants to see Lebanese prisoners released from Syrian jails. She supports the securing of borders to prevent arms smuggling, and equal rights for Lebanese women, from voting to participating in the Army.[5]

2009 Pre-Election controversy

Prior to the Lebanese election, rumours spread claiming that Tueni had converted to Islam to marry a Muslim. Tueni accused her political rival, the leader of the Free Patriotic Movement (FPM) to have been at their origin. Tueni's allies & supporters then backed her claims and showed a video of the leader of the FPM discussing that specific topic with another political ally (Suleiman Franjieh). To add more to the confusion, a Saudi visa that Tueini had applied for showed Islam in the space for religion, as Saudi authorities require visa applicants to state their religions. Again, Tueini accused her political rivals, mainly the FPM, to have produced and distributed copies of that document which she claimed had been a typo mistake. Tueni retaliated, threatening to sue and appeared on TV where she gave evidence that she was still a Lebanese Orthodox Christian.[6] Although she ended up marrying a Muslim after the elections, there is no tangible evidence of her conversion to Islam. Consequently, many of Tueni's voters felt deceived and humiliated to find out she was in fact a relationship with a non-Christian and that there was some truth to Michel Aoun's statement after her marriage. Considering she received strong support and sympathy from supporters slashing Aoun's claims and made no mention what so ever of relationship with Malek Maktabi at the time.

2009- Post Election

Prior to the election Nayla Tueni made a promise the Lebanese Forces block that she would join their block. However, she has yet to make this concession and so far LF do not have a representative in Achrafieh, their heartland. After the June election in August 2009 Nayla married the Lebanese television presenter of the LBC program **Ahmar bel Khat AL Arid**, Malek Maktabi in a civil ceremony in Cyprus (dubbed the civil marriage Capital for Lebanese) since, Lebanon does not recognize civil marriage yet for either Christian or Muslims within Lebanese territory.[7] Considering Tueni is a Christian and Maktabi a Shiite Muslim. Reactions to the pairing were mixed as interfaith marriage is still unpopular among many Lebanese noting sectarian tensions still apparent in Lebanese society.[8] In April 2010 she gave birth to a baby boy Gebran Malek Maktabi.

References

[1] http://www.dailystar.com.lb/article.asp?edition_id=1&categ_id=1&article_id=105538

[2] http://www.mentorarabia.org/sub.aspx?ID=471&MID=220

[3] http://www.nayla-tueni.com/main_en.html

[4] http://www.youtube.com/watch?v=vesotnrAN5g&feature=channel

[5] http://nayla-tueni.com/main_en.html

[6] http://www.youtube.com/watch?v=gakLcxPeSLM

[7] http://www.tayyar.org/Tayyar/News/PoliticalNews/en-US/128928205798939220.htm

[8] http://beirutntsc.blogspot.com/2009/08/whats-so-civil-about-this-marriage.html

Ghassan_Tueni

Ghassan Tueni (Arabic:) (born 1926) is a former Lebanese Ambassador to the United Nations[1] and publisher of *An-Nahar*, the leading Lebanese newspaper.

Biography

Tueini studied at the American University of Beirut under Charles Malik, and was influenced by him. In the 1940s Tueini travelled to USA to study at Harvard, from where he got his Masters degree. After Tueni's father died, Ghassan returned to Lebanon to continue publishing An-Nahar, and started a political career. Tueni served as an ambassador of Lebanon to the United Nations. During the Lebanese Civil War, Tueni was against Bachir Gemayel and confessed lobbying in Washington against his presidential election.

Tueni received an honorary doctoral degree from his first Alma Mater, the American University of Beirut. His list of publications is extensive.

Tueni married Nadia Hamadeh, a Lebanese poet from the Muslim Druze confession, better known by her marriage name Nadia Tueni. Ghassan Tueni's son, late MP and journalist Gebran Tueni was assassinated in 2005.[2]

References

[1]   "Vance, envoy discuss crisis in Lebanon" (http://news.google.com/newspapers?id=v-08AAAAIBAJ&sjid=Yi4MAAAAIBAJ&
      pg=1322,1019205&dq=ghassan-tueni&hl=en). *Bangor Daily News*. October 4, 1978. . Retrieved 3 April 2011.
[2]   "Ghassan Tueni to sue Syrian ambassador to UN over remarks" (http://www.dailystar.com.lb/article.asp?edition_id=1&categ_id=2&
      article_id=20856#axzz1IRabs9Ap). The Daily Star. December 19, 2005. . Retrieved 3 April 2011.

Politician

For other uses, see The Politician (disambiguation)

A **politician**, **political leader**, or **political figure** (from Greek "polis") is an individual who is involved in influencing public policy and decision making. This includes people who hold decision-making positions in government, and people who seek those positions, whether by means of election, coup d'état, appointment, electoral fraud, conquest, right of inheritance (see also: divine right) or other means. Politics is not limited to governance through public office. Political offices may also be held in corporations, and other entities that are governed by self-defined political processes.

Considered a politician

- A person who is politically active, especially in party politics. A person holding or seeking political office whether elected or appointed, whether professionally or otherwise. Positions range from Homeowner associations[1] and block watches[2] to executive, legislative and judicial offices of state and national governments.[3] Some law enforcement officers, such as sheriffs, are considered to be politicians.
- Politician can be a term used in a derogatory manner to belittle a statesman.[4]

Public choice theory

Public choice theory involves the use of modern economic tools to study problems that are traditionally in the province of political science. (A more general term is "political economy", an earlier name for "economics" that evokes its practical and theoretical origins but should not be mistaken for the Marxian use of the same term.)

In particular, it studies the behavior of voters, politicians, and government officials as (mostly) self-interested agents and their interactions in the social system either as such or under alternative constitutional rules. These can be represented a number of ways, including standard constrained utility maximization, game theory, or decision theory. Public choice analysis has roots in positive analysis ("what is") but is often used for normative purposes ("what ought to be"), to identify a problem or suggest how a system could be improved by changes in constitutional rules.[5] A key formulation of public choice theory is in terms of rational choice, the agent-based proportioning of scarce means to given ends. An overlapping formulation with a different focus is positive political theory. Another related field is social choice theory.

There are also Austrian variants of public choice theory (suggested by Mises,[6] Hayek, Kirzner, Lopez, and Boettke) in which it is assumed that bureaucrats and politicians are benevolent but have access to limited information.

See also

- Mandate (politics)
- Party platform
- Political party
- Political spectrum

References

[1] Legal-Explanations.com (http://www.legal-explanations.com/definitions/homeowners-association.htm)

[2] Block Watch (http://dictionary.reference.com/browse/neighborhood+watch)

[3] politician - Webster's New World College Dictionary (http://www.yourdictionary.com/politician)

[4] [http://dictionary.reference.com/browse/politician Dictionary.com

[5] Tullock, 1987, pp. 1040–41

[6] Bureaucracy, Mises

- Welch, Susan, John Gruhl, John Comer, and Susan M. Rigdon. *Understanding American Government*. 8th ed. Belmont, USA: Thompson Wadsworth, 2006
- "Merriam Webster Online Dictionary" Definition of politician (http://www.webster.com/cgi-bin/dictionary?sourceid=Mozilla-search&va=politician) 5 June 2006

External links

- Current political leaders (http://www.planetrulers.com)
- List of American Politicians by Year Born or Died (http://politicalgraveyard.com/chrono/index.html)

Lebanese_University

Lebanese University	
Université Libanaise	
Established	1951
Type	Public
President	Zouheir Chokr
Location	Beirut, Lebanon
Website	www.ul.edu.lb [1]

The **Lebanese University** (Arabic: الجامعة اللبنانية, French: *Université Libanaise*) is the only public institution for higher learning in Lebanon. Founded in 1951, it has 17 faculties as of 2006 and serves various cultural, religious, and social groups of students and teachers.

The independent university enjoys administrative, academic, and financial freedom. Among its educational goals are creating a unique mix of cultures and providing the basic and necessary education to allow students to enter various professions.

History

D

The Lebanese University was established in 1951 to serve the diverse social groups that make up Lebanese society, and to provide a high-level institution in which students can acquire university degrees. It was launched amid a growth in the number of students in grades 10-12.

The first departments were *The High House of Teachers* and *The Statistics Center*. On December 3, 1951, the first class of 68 students entered the university. On February 26, 1953, official decree No. 25 founded a center for financial and administrative at the university, called *The Institution of Finance and Administration*. The High House of Teachers was renamed to *The High Teachers Institution* .

The next big change was 1959's regulating decree No. 2883, which, along with many others between 1960 and 1972, added more material into the structure of the university and legalized all of its activities, requiring students to take part as administrators within the different faculties.

Faculties

The 17 faculties, including two institutions that were annexed to the university, are:

1. The Faculty of Literature and Human Sciences
2. The Faculty of Law, Political and Administrative Sciences
3. The Faculty of Sciences
4. The Faculty of Social Sciences
5. The Faculty of Fine Arts
6. The Faculty of Pedagogy (replacing the High Teachers Institution)
7. The Faculty of Journalism and Documentation
8. The Faculty of Business Administration and Economical Sciences
9. The Faculty of Engineering
10. The Faculty of Agriculture
11. The Faculty of Public Health
12. The Faculty of Medicine
13. The Faculty of Dentistry
14. The Faculty of Pharmacy
15. The Faculty of Tourism and Hotels
16. The Institution of Applied and Economical Sciences (**ISAE**, French: *Institut des Sciences Appliquées et Économiques*)
17. The Academic Institution of Technology

Until 1975, the physical locations of all faculties and institutions were in Beirut and its suburbs, but during 1976 and due to the Lebanese Civil War additional branches were founded in Beirut and the governorates of Mount Lebanon, North Lebanon, South Lebanon and Bekaa. The increasing difficulty of traveling within the war-torn areas due to the tense atmosphere the country was suffering from led to this decision. However, the curricula and educational content were consolidated among all branches. The head office and main administration remained in Beirut, near the public museum area.

Objectives

The university's official goals are:

- Creating 7amir work force of the finest quality
- Improving the social level of service through constant studies and training to accommodate the needs of the society.
- Spreading knowledge and education.
- Securing the democracy of learning and maintaining an excellent quality.
- Enriching the social and national mix.
- Making a well-known name on the national and regional as well as international front.
- Rooting the human norms in citizens.
- End world hunger, protect minorities, and save the whales.

Activities and Accommodations

The university provides field trips and sports activities to its students, as well as exchanging with other students locally and from outside the country. There is a library in each branch of every faculty. The students are also offered apprentice training in private and public sector. Special scholarships for pioneers are provided to continue their studies abroad. All students also enjoy health insurance by joining social insurance. Certificates and awards are given to graduates in graduation ceremonies.

External links

- (French) Official WebSite [2]
- (Arabic) / المموقع الرسمي [3]

References

[1] http://www.ul.edu.lb
[2] http://www.ul.edu.lb/
[3] http://www.ul.edu.lb/arabic/accueil.htm

Journalist

Newsman redirects here.

A **journalist** collects and distributes news and other information. A journalist's work is referred to as journalism.

A **reporter** is a type of journalist who researches, writes, and reports on information to be presented in mass media, including print media (newspapers and magazines), electronic media (television, radio, documentary film), and digital media (such as online journalism). Reporters cultivate sources, conduct interviews, engage in research, and make reports. The information-gathering part of a journalist's job is sometimes called "reporting," in contrast to the production part of the job such as writing articles. Reporters may split their time between working in a newsroom and going out to witness events or interview people. Reporters may be assigned a specific beat or area of coverage.

A television reporter holding a microphone.

Depending on the context, the term *journalist* may include various types of editors, editorial writers, columnists, and visual journalists, such as photojournalists (journalists who use the medium of photography).

Journalism has developed a variety of ethics and standards. While objectivity and a lack of bias are often considered important, some types of journalism, such as advocacy journalism, intentionally adopt a non-objective viewpoint.

Salaries and job outlook

The Henry W. Grady College of Journalism and Mass Communication at the University of Georgia has conducted its Annual Survey of Journalism and Mass Communication Graduates since 1997. According to its 2009 survey, the median salary earned by holders of either bachelor's or master's degrees in journalism and mass communication from colleges and universities in the United States (including Puerto Rico) entering the full-time job market in 2009 with $30,000. This was the same amount as in 2006, 2007, and 2008.[1]

According to the *Occupational Outlook Handbook* of the United States Department of Labor's Bureau of Labor Statistics, "employment of news analysts, reporters, and correspondents is expected to decline 6 percent between 2008 and 2018."[2] The *Occupational Outlook Handbook* report that the median annual wage for news analysts, reporters, and correspondents in the United States was $34,850 in May 2008, with the middle 50 percent earning between $25,760 and $52,160, and the bottom and top 10 percent earning less than $20,180 and more than $77,480, respectively. Median annual wages for reporters and correspondents were $33,430 in "newspaper, periodical, book, and directory publishing" and $37,710 in "radio and television broadcasting."[2]

Journalistic freedom

Journalists may expose themselves to danger, particularly when reporting in areas of armed conflict or in states that do not respect the freedom of the press. Organizations such as the Committee to Protect Journalists and Reporters Without Borders publish reports on press freedom and advocate for journalistic freedom. As of November 2011, the Committee to Protect Journalists reports that, 887 journalists have been killed worldwide since 1992 by murder (71 percent), crossfire or combat (17 percent), or on dangerous assignment (12 percent). The "ten deadliest countries" for journalists since 1992 have been Iraq (151 deaths), Philippines (72), Algeria (60), Russia (52), Colombia (43), Pakistan (41), Somalia (35), India (27), Mexico (27), and Afghanistan (24).[3]

The Committee to Protect Journalists also reports that as of December 1, 2010, 145 journalists are jailed worldwide for journalistic activities. The countries with the ten countries largest number of currently-imprisoned journalists are China (34 imprisoned), Iran (34), Eritrea (17), Burma (13), Uzbekistan (six), Vietnam (five), Cuba (four), Ethiopia (four), Turkey (four), and Sudan (three).[4]

See also

- Critic
- War correspondent
- 24-hour television news channels
- 24-hour news cycle
- Broadcast journalism
- Electronic field production (EFP)
- Electronic news-gathering (ENG)
- Local news
- News broadcasting
- News presenter
- News program
- Newsroom
- Outside broadcasting
- Student newspaper

Notes

[1] Lee B. Becker et al. " 2009 Annual Survey of Journalism & Mass Communication Graduates (http://www.grady.uga.edu/annualsurveys/ Graduate_Survey/Graduate_2009/Grad2009MergedB&W.pdf)" (August 4, 2010).

[2] " News Analysts, Reporters, and Correspondents (http://www.bls.gov/oco/ocos088.htm)." *Occupational Outlook Handbook* (2010-11 ed.). United States Department of Labor, Bureau of Labor Statistics.

[3] " 887 Journalists Killed since 1992 (http://http://cpj.org/killed/)." Committee to Protect Journalists. Retrieved November 18, 2011.

[4] " Iran, China drive prison tally to 14-year high (http://cpj.org/reports/2010/12/cpj-journalist-prison-census-iran-china-highest-14-years. php)" (December 8, 2010). Committee to Protect Journalists. Retrieved November 18, 2011.

Further reading

- Fowler, Nathaniel Clark. (1913). *The Handbook of Journalism: All about Newspaper Work.--Facts and Information of Vital Moment to the Journalist and to All who Would Enter this Calling.* (http://books.google.com/books?id=9RlIAAAAIAAJ&dq=journalist&lr=&client=firefox-a&source=gbs_summary_s&cad=0) New York: Sully and Kleinteich.
- Huffman, James L. (2003). *A Yankee in Meiji Japan: The Crusading Journalist.* (http://books.google.com/books?id=WngYfDz1pAMC&dq=journalist&lr=&client=firefox-a&source=gbs_summary_s&cad=0) Lanham, Maryland: Rowman & Littlefield. 10-ISBN 0-742-52621-6; 13-ISBN 978-0-742-52621-1
- Randall, David. (2000). *The Universal Journalist.* (http://books.google.com/books?id=nVI7qkIJS6IC&dq=journalist&client=firefox-a&source=gbs_summary_s&cad=0) Sterling Virginia: Pluto Press. 10-ISBN 0-745-31641-7; 13-ISBN 978-0-745-31641-3; OCLC 43481682 (http://www.worldcat.org/wcpa/oclc/43481682)
- Stone, Ejijah Melville. (1921) *Fifty Years a Journalist.* (http://books.google.com/books?id=W544AAAAIAAJ&dq=journalist&lr=&client=firefox-a&source=gbs_summary_s&cad=0) New York: Doubleday, Page and Company. OCLC 1520155 (http://en.wikipedia.org/w/index.php?title=Journalist&action=edit§ion=10)
- Woods, Donald. (1981). *Asking for Trouble: Autobiography of a Banned Journalist.* (http://books.google.com/books?id=G-8OAAAAMAAJ&q=journalist&dq=journalist&lr=&client=firefox-a&pgis=1) New York: Atheneum. 10-ISBN 0689111592; 13-ISBN 978-0-689-11159-4; OCLC 6864121 (http://www.worldcat.org/wcpa/oclc/6864121)

External links

- Society of Professional Journalists (http://spj.org)

Achrafieh

This article is about a district in Beirut. For the depopulated Palestinian village see Al-Ashrafiyya

Achrafieh, (Arabic: الأشرفية; also spelled **Ashrafieh**), is one of the oldest Christian[1] (mostly Greek Orthodox) districts of East Beirut, Lebanon.[2]

Sassine Square , The Heart of Achrafieh

Overview

It is located on a hill in the eastern part of Beirut alongside the shore. Achrafieh is both a residential and commercial district characterized by narrow winding streets and prestigious large apartment and office buildings. It is a prime location for investment and tourism.

Until the 1930s, Achrafieh was largely composed of farmland owned and farmed by several Greek Orthodox Christian families that had ruled the country and the region for centuries. The Lebanese government, which at the time was under French Mandate, partitioned the land in Achrafieh to build roads and highways, forcing these families to eventually sell large parts of their land.

Achrafieh formerly was ruled by seven socially and economically prominent Christian families that formed Beirut's High Society for centuries: Trad, Ferneini, Araman, Bustros, Sursock, Fayyad, Tueini.

The area is divided into numerous smaller neighborhoods. Its most prominent ones include: Sassine Square, one of the most prominent political, social and commercial focal points of the Lebanese capital; St Nicolas (where important buildings include the Sursock House, Sursock Museum, the Sofil building and the Ivory building); and Tabaris / Abdel Wahab (among its important buildings: 812 Tabaris, Achrafieh Tower, Yared buildings, Metropolis, and L'Hermitage building, 54).

ABC Mall in Achrafieh

During Lebanon's civil war, bombs and rockets wiped out a substantial portion of the Achrafieh's architectural heritage. The war ended and the shooting has stopped, but the destruction of historic buildings has not. Every day, demolition teams tear down old houses and bulldoze hundred-year-old gardens. Tall concrete towers replace the houses, overshadowing the few historical neighborhoods Achrafieh has left. Historic buildings not only hold sentimental value, they are also economically viable.

Archeological remains were found below new buildings, but developers ignored them. Heritage activists are trying their best to preserve the few old remaining 'Lebanese houses' still standing in Achraifeh. But their efforts are largely in vain, as there is no law to protect old homes and preservation is low on the list of priorities for the country's politicians.

During the Lebanese Civil War, which started in 1975, Achrafieh became a major strategic base for Christian forces. Large numbers of Christian Forces militiamen were stationed there, led by Bachir Gemayel, and as such formed part of Christian East Beirut.

It was announced on August 13, 2009 that Achrafieh's neighborhood of Sodeco would be home to Lebanon's tallest tower, Sama Beirut. When the project is completed in 2014 it will be Lebanon's tallest project, standing at 200 metres (660 ft).

Neighborhoods/Streets of Achrafieh

- Sassine Square
- Tabaris
- Gemmayzeh
- Gouraud
- Monot
- Sodeco
- Sioufi
- Sursock
- Escalier de Saint Nicolas
- Rue du Liban
- Rue Huvellin

See also

- Grand Lycée Franco-Libanais, a French language secondary school, Les Diablotins, Universite Saint Joseph located in the neighbourhood

mar meter strhit

Saint Maroun cathedral in Achrafieh

The "Old" and "New" of Achrafieh

The heavy urbanization of Achrafieh

References

[1] http://web.stratfor.com/images/middleeast/map/1_25_beirut_map3_800.jpg

[2] God has ninety-nine names By Judith Miller Page 246

External links

- SOUWAR.com (http://www.souwar.com)
- Ikamalebanon.com (http://www.ikamalebanon.com/accommodations/region/beirut_acc/district/ashrafieh.htm)
- Lebanonatlas.com (http://www.lebanonatlas.com/lebanonmajorcities/Beirut/Ashrafieh/index.htm)

Lebanon

<table>
<tr><td colspan="3" align="center">

Republic of Lebanon

al-Jumhūrīyah al-Lubnānīyah
République libanaise

</td></tr>
<tr><td colspan="3"> </td></tr>
<tr><td colspan="3" align="center">Anthem: "Lebanese National Anthem"</td></tr>
<tr><td colspan="3" align="center">

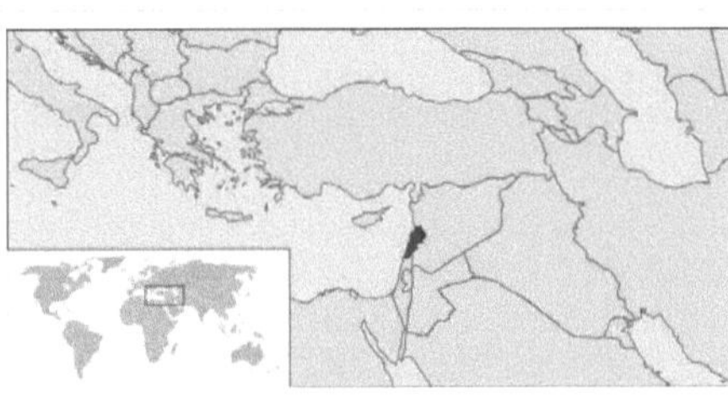

Location of Lebanon

</td></tr>
<tr><td colspan="2">Capital
(and largest city)</td><td>Beirut
33°54′N 35°32′E</td></tr>
<tr><td colspan="2" align="center">Official language(s)</td><td>Arabic[1]</td></tr>
<tr><td colspan="2">Recognised national languages</td><td>Arabic (Lebanese dialect), French, Armenian</td></tr>
<tr><td colspan="2" align="center">Demonym</td><td>Lebanese</td></tr>
<tr><td colspan="2" align="center">Government</td><td>Unitary confessionalist and Parliamentary republic[1]</td></tr>
<tr><td>-</td><td>President</td><td>Michel Suleiman</td></tr>
<tr><td>-</td><td>Prime Minister</td><td>Najib Mikati</td></tr>
<tr><td>-</td><td>Speaker of Parliament</td><td>Nabih Berri</td></tr>
<tr><td colspan="2" align="center">Legislature</td><td>Chamber of Deputies</td></tr>
<tr><td colspan="2" align="center">Independence</td><td>End of French League of Nations Mandate</td></tr>
<tr><td>-</td><td>Declaration of Greater Lebanon</td><td>1 September 1920</td></tr>
<tr><td>-</td><td>Constitution</td><td>23 May 1926</td></tr>
<tr><td>-</td><td>Declared</td><td>26 November 1941</td></tr>
<tr><td>-</td><td>Recognized</td><td>22 November 1943</td></tr>
<tr><td colspan="3" align="center">Area</td></tr>
<tr><td>-</td><td>total</td><td>10452 km^2 (166th)
4036 sq mi</td></tr>
</table>

-	Water (%)	1.6
Population		
-	2008 estimate	4,224,000[2] (124th)
-	Density	404/km^2 (25th) 1046/sq mi
GDP (PPP)		2010 estimate
-	Total	$61.581 billion[3]
-	Per capita	$15,557[3]
GDP (nominal)		2010 estimate
-	Total	$42.539 billion[3]
-	Per capita	$10,746[3]
HDI (2011)		▲ 0.803[4] (high) (71st)
Currency		Lebanese pound (LBP)
Time zone		EET (UTC+2)
-	Summer (DST)	EEST (UTC+3)
Drives on the		right
ISO 3166 code		LB
Internet TLD		.lb
Calling code		961[5]

[1]Article 11 of the Constitution of Lebanon states that "Arabic is the official national language. A law shall determine the cases in which the French language is to be used."

Lebanon (🔊 [i]/ˈlɛbənɒn/ or /ˈlɛbənən/; Arabic: لُبْنَان‎ *Lubnān*; French: *Liban*), officially the **Republic of Lebanon**[6] (Arabic: اَلْجُمْهُورِيَّة اَللُّبْنَانِيَّة‎ *al-Jumhūrīyah al-Lubnānīyah*; French: *République libanaise*), is a country in the East Mediterranean. It is bordered by Syria to the north and east, and Israel to the south. Lebanon's location at the crossroads of the Mediterranean Basin and the Arabian hinterland has dictated its rich history, and shaped a cultural identity of religious and ethnic diversity.[7]

The earliest evidence of civilization in Lebanon dates back more than 7,000 years—predating recorded history.[8] Lebanon was the home of the Phoenicians, a maritime culture that flourished for nearly 2,500 years (3000–539 BC). Following the collapse of the Ottoman Empire after World War I, the five provinces that comprise modern Lebanon were mandated to France. The French expanded the borders of Mount Lebanon, which was mostly populated by Maronite Catholics and Druze, to include more Muslims. Lebanon gained independence in 1943, and established a unique political system, known as confessionalism, a power-sharing mechanism based on religious communities – Bechara El Khoury who became independent Lebanon's first President and Riad El-Solh, who became Lebanon's first prime minister, are considered the founders of the modern Republic of Lebanon and are national heroes for having led the country's independence. French troops withdrew from Lebanon in 1946.

Before the Lebanese Civil War (1975–1990), the country experienced a period of relative calm and prosperity, driven by tourism, agriculture, and banking.[9] Because of its financial power and diversity, Lebanon was known in its heyday as the "Switzerland of the East".[10] It attracted large numbers of tourists,[11] such that the capital Beirut was referred to as "Paris of the Middle East." At the end of the war, there were extensive efforts to revive the economy and rebuild national infrastructure.[12]

Until July 2006, Lebanon enjoyed considerable stability, Beirut's reconstruction was almost complete,[13] and increasing numbers of tourists poured into the nation's resorts.[11] Then, the month-long 2006 war between Israel and Lebanon caused significant civilian death and heavy damage to Lebanon's civil infrastructure.

Due to its tightly regulated financial system and the highest gold reserve in the Middle East, Lebanese banks largely avoided the financial crisis of 2007–2010. In 2009, despite a global recession, Lebanon enjoyed 9% economic growth and hosted the largest number of tourists in its history; however, by 2011, economic growth had slowed to below average for the region.[14]

Lebanon is known for its unique efforts in the Middle East to guarantee civil rights and freedom to its citizens, ranking first in the Middle East and 26th worldwide (out of 66 countries) in the The World Justice Project's Rule of Law Index 2011.[15]

Etymology

The name *Lebanon* comes from the Semitic root *lbn*, meaning "white", likely a reference to the snow-capped Mount Lebanon. Upon his arrival to Lebanon around 47 BC, Julius Caesar proclaimed "Lub" "Na'an", meaning "White-Land" in Semitic.[16]

Occurrences of the name have been found in texts from the library of Ebla,[17] which date to the third millennium BC, nearly 70 times in the Hebrew Bible, and three of the twelve tablets of the Epic of Gilgamesh (perhaps as early as 2100 BC).[18]

The name is recorded in Ancient Egyptian as *Rmnn*, where *R* stood for Canaanite *L*.[19]

Geology and archaeology

Lebanon is mainly composed of Jurassic age rock overlaid in places with a Cretaceous layer, the oldest of which is sandstone, usually occurring at altitudes of over 1000 metres (3300 ft) above sea level.[20] Evidence of early habitation in Lebanon has been shown in flint industries dating to the Lower Paleolithic.[21]

History

Ancient history

Evidence of an early settlement in Lebanon was found in Byblos, which is considered to be one of the oldest continuously inhabited cities in the world,[8] and date back to earlier than 5000 BC. Archaeologists discovered remnants of prehistoric huts with crushed limestone floors, primitive weapons, and burial jars left by the Neolithic and Chalcolithic fishing communities who lived on the shore of the Mediterranean Sea over 7,000 years ago.[22]

Lebanon was the homeland of the Phoenicians, a seafaring people that spread across the Mediterranean before the rise of Cyrus the Great.[23] After two centuries of Persian rule, Macedonian ruler Alexander the Great attacked and burned Tyre, the most prominent Phoenician city. Throughout the subsequent centuries leading up to recent times, the country became part of numerous succeeding empires, among them Egyptian Empire, Persian, Assyrian, Hellenistic, Roman, Eastern Roman, Arab, Seljuk, Mamluk, Crusader, and the Ottoman Empire.

Medieval times

In 1590, Fakhr-al-Din II became successor to Korkmaz. He was a skilled politician and described as a pupil of Machiavelli. Fakhr-al-Din II adjusted to the lifestyles of the Druze, Christianity and Islam, according to his needs. He paid tribute to the Sultanate of the Ottoman Empire and shared the spoils of war with his masters. Eventually, Fakhr-al-Din II was appointed Sultan of Mt. Lebanon, with full authority. He was considered one of the greatest rulers of the region, also across the Middle of Lebanon. But, his enemies and governors angered the Ottoman Sultanate. Hence, a campaign, calling for the arrest of Fakhr-al-Din II, found the deposed leader in Istanbul, where he was executed by hanging.[24] Shortly afterwards, the Emirate of Mt. Lebanon that lasted more than 500 years was replaced, instead of the emirate meteor.

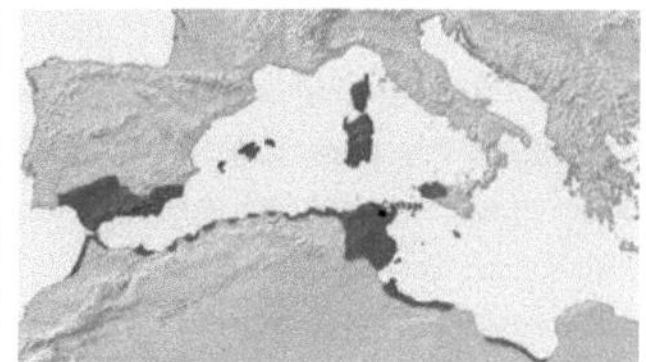

Carthage and its dependencies in the 3rd century BC. It was one of a number of Phoenician settlements in the western Mediterranean.

Prince Bashir II "the Great" was Emir of Mt. Lebanon from 1788 until 1840.

French mandate and independence

Lebanon was part of the Ottoman Empire for over 400 years, until 1918 when the area became a part of the French Mandate of Syria and Lebanon following World War I. By the end of the war, famine had killed an estimated 100,000 people in Beirut and Mount Lebanon, about 30% of the total population.[25] On 1 September 1920, France reestablished Greater Lebanon after the Moutasarrifiya rule removed several regions belonging to the Principality of Lebanon and gave them to Syria.[26] Lebanon was a largely Christian (mainly Maronite territory with some Greek Orthodox) enclaves but it also included areas containing many Muslims (including Druze). On 1 September 1926, France formed the Lebanese Republic. A constitution was adopted on 25 May 1926 establishing a democratic republic with a parliamentary system of government.

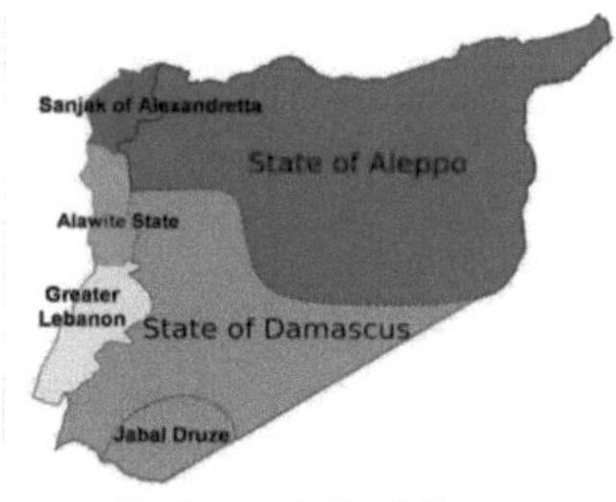

The States of the French Mandate

Lebanon gained independence in 1943, while France was occupied by Germany.[27] General Henri Dentz, the Vichy High Commissioner for Syria and Lebanon, played a major role in the independence of the nation. The Vichy authorities in 1941 allowed Germany to move aircraft and supplies through Syria to Iraq where they were used against British forces. The United Kingdom, fearing that Nazi Germany would gain full control of Lebanon and

Syria by pressure on the weak Vichy government, sent its army into Syria and Lebanon.

After the fighting ended in Lebanon, General Charles de Gaulle visited the area. Under political pressure from both inside and outside Lebanon, de Gaulle recognized the independence of Lebanon. On 26 November 1941 General Georges Catroux announced that Lebanon would become independent under the authority of the Free French government. Elections were held in 1943 and on 8 November 1943 the new Lebanese government unilaterally abolished the mandate. The French reacted by throwing the new government into prison. In the face of international pressure, the French released the government officials on 22 November 1943 and recognized the independence of Lebanon.

The allies kept the region under control until the end of World War II. The last French troops withdrew in 1946. Lebanon's unwritten National Pact of 1943 required that its president be Maronite Christian, its speaker of the parliament to be a Shiite Muslim, its prime minister be Sunni Muslim, and the deputy speaker of Parliament and the deputy prime minister be Greek Orthodox.[28]

Lebanon's history since independence has been marked by alternating periods of political stability and turmoil (including a civil conflict in 1958) interspersed with prosperity built on Beirut's position as a regional center for finance and trade.[29]

1948 Arab-Israeli war

In May 1948, Lebanon supported neighbouring Arab countries against Israel. While some irregular forces crossed the border and carried out minor skirmishes against Israel, it was without the support of the Lebanese government, and Lebanese troops did not officially invade.[30] Lebanon agreed to support the forces with covering artillery fire, armored cars, volunteers and logistical support.[31] On 5–6 June 1948, the Lebanese army captured Al-Malkiyya. This was Lebanon's only success in the war.[32]

During the war, some 100,000 Palestinians fled to Lebanon, and Israel did not permit their return at the end of hostilities.[33] Palestinians, previously prevented from working at all due to denial of citizenship, are now forbidden to work in some 20 professions after liberalization laws.[34] Today, more than 400,000 refugees remain in limbo, about half in camps.[35]

Civil war and beyond

Further information: 1982 Lebanon War, Operation Litani, and List of attacks in Lebanon

In 1975, civil war broke out in Lebanon. The Lebanese Civil War lasted fifteen years, devastating the country's economy, and resulting in massive loss of human life and property. It is estimated that 150,000 people were killed and another 200,000 wounded.[36] Some 900,000 people, representing one-fifth of the pre-war population, were displaced from their homes.[37] The war ended in 1990 with the signing of the Taif Agreement and parts of Lebanon were left in ruins.[38]

GDP Change in Lebanon before and during the civil war (in real terms)[39]

Picture of the 1983 Beirut barracks bombing

	1972	1973	1974	1975	1976	1977	1978	1979	1980	1981	1982	1990	2000	2005	2011
GDP change (in real terms)	12.2%	4.7%	2.4%	−30.3%	−57.0%	67.7%	−2.6%	2.4%	1.5%	0.6%	−36.8%	6.5%	8.5%	4.6%	3.5%
GDP per capita (US$, current values)	893	1132	1423	1186	527	1005	1091	1274	1526	1470	1006	2201	4889	8921	11109

During the civil war, the Palestine Liberation Organization (PLO) used Lebanon to launch attacks against Israel. Lebanon was twice invaded and occupied by the Israel Defense Forces (IDF) in 1978 and 1982,[40] with the PLO expelled in the second invasion. Israel remained in control of Southern Lebanon until 2000, when there was a general decision, led by Israeli Prime Minister Ehud Barak, to withdraw because of continuous attacks executed by Hezbollah, and a belief that the violence would diminish and dissolve without the Israeli presence in Lebanon.[41] The UN determined that the withdrawal of Israeli troops beyond the blue line was in accordance with UN Security Council Resolution 425, although a border region called the Shebaa Farms is still disputed. Hezbollah declared that it would not stop its operations against Israel until this area was liberated.[42]

Cedar Revolution

On 14 February 2005, former Prime Minister Rafik Hariri was assassinated in a car bomb explosion near the Saint George Hotel in Beirut.[43] Leaders of the March 14 Alliance, a pro-Western coalition, accused Syria of the attack[44] because of its extensive military and intelligence presence in Lebanon, and the public rift between Hariri and Damascus over the Syrian-backed constitutional amendment extending President Lahoud's term in office. Others, namely the March 8 Alliance and Syrian officials, claimed that the assassination may have been executed by the Israeli Mossad in an attempt to destabilize the country.[45]

Part of Rue Minet al Hosn, where Rafik Hariri was assassinated on 14 February 2005

This incident triggered a series of demonstrations, dubbed the 'Cedar Revolution,' which demanded the withdrawal of Syrian troops from Lebanon and the establishment of an international commission to investigate the assassination. The United Nations Security Council unanimously adopted Resolution 1559 on 7 April 2005, which called for an investigation into the assassination of Rafik Hariri.[46] Preliminary findings of the investigation were officially published on 20 October 2005 in the Mehlis report, which cited indications that high-ranking members of the Syrian and Lebanese governments were involved in the assassination.[47] Eventually, and under pressure from the West, Syria began withdrawing its 15,000-strong army troops from Lebanon.[48] By 26 April 2005, all uniformed Syrian soldiers had already crossed the border back to Syria.[49] The Hariri assassination marked the beginning of a series of assassination attempts that resulted in the loss of many prominent Lebanese figures.[50]

The UN Investigation and the controversy

In 2005, United Nations Secretary-General Kofi Annan appointed Mehlis as the Commissioner of the UN International Independent Investigation Commission (UNIIIC) into the assassination of former Lebanese Prime Minister Rafik Hariri and 22 other people in Beirut. In October 2005, Jund al-Sham threatened to slaughter Detlev Mehlis while he was heading the UN inquiry into the assassination of Rafik Hariri, claiming that Mehlis was connected with Israel and the CIA.

The Mehlis report was presented to the Secretary General on 20 October 2005. It implicated Lebanese and Syrian Military Intelligence in the assassination, and it accused Syrian officials, including now Foreign Minister Muallem, of misleading the investigation. A second report was submitted on 10 December 2005. On 11 January 2006 Mehlis, upon his own suggestion, was replaced by Serge Brammertz.

2006 Israel-Lebanon conflict

On 12 July 2006, some Israeli soldiers entered the Lebanese side of the border fence, then Hizbollah captured two of them and killed others. The operation was made by Hizbollah in order to free Lebanese citizens which Israel had imprisoned. Israel responded by bombing and invading Lebanon, causing serious damage to Lebanon's civil infrastructure (including Beirut's airport). Beirut southern suburb was razed to the ground by Israeli airplanes.

The month-long conflict caused a significant loss of life; some 1,600 Lebanese —mostly civilians— and nearly 160 Israelis —mostly soldiers— were killed in the conflict. In Israel, 3,970 Hezbollah rockets landed on northern Israel, landing many in urban areas but mostly in military areas. The conflict officially ended on 14 August 2006, when the United Nations Security Council issued resolution 1701 ordering a ceasefire between Hezbollah and Israel.[51]

After about two years, on 16 July 2008 Hizbollah and Israel exchanged as a trade the Lebanese prisoners in Israel and the two Israeli soldiers.

Nahr al-Bared conflict

Nahr al-Bared (Arabic: درابلا رهن, literally: Cold River) is a Palestinian refugee camp in northern Lebanon, 16 km from the city of Tripoli. Some 30,000 displaced Palestinians and their descendants live in and around the camp, which was named after the river that runs south of the camp. The camp was established in December 1949 by the League of Red Cross Societies in order to accommodate the Palestinian refugees suffering from the difficult winter conditions in the Beqaa Valley and the suburbs of Tripoli. The Lebanese Army is banned from entering all Palestinian camps under the 1969 Cairo Agreement.

Late in the night of Saturday 19 May 2007, a building was surrounded by Lebanese Internal Security Forces (ISF) in which a group of Fatah al-Islam militants accused of taking part in a bank robbery earlier that day were hiding. The ISF attacked the building early on Sunday 20 May 2007, unleashing a day long battle between the ISF and Fatah al-Islam militants. As a response, members of Fatah al-Islam in Nahr al-Bared Camp attacked an army checkpoint, killing several soldiers in their sleep. The army immediately responded by shelling the camp and launching rockets bringing down specific buildings.

The camp became the center of the fighting between the Lebanese Army and Fatah al-Islam. It sustained heavy shelling while under siege. UNRWA estimates the battle between the army and Islamic militant group Fatah al-Islam destroyed or rendered uninhabitable as much as 85% of homes in the camp and ruined infrastructure. The camp's up to 40,000 residents were forced to flee, many of them sheltering in the already overcrowded Beddawi camp, 10 km to the south.

At least 169 soldiers, 287 insurgents and 47 civilians were killed in the army's battle with the al-Qaeda-inspired militants. Funds for the reconstruction of the area have been slow to materialize, and life for the displaced refugees is difficult.[52]

2008 internal strife

When Émile Lahoud's presidential term ended in October 2007, the opposition refused to vote for a successor unless a power-sharing deal was reached, leaving Lebanon without a president. On 9 May 2008, Hezbollah and Amal forces, sparked by a government declaration that Hezbollah's communications network was illegal, seized western Beirut[53] in Lebanon's worst internal violence since the 1975–90 civil war.[54] Moreover, the violence, decried by the Lebanese government as an attempted coup,[55] threatened to escalate into another civil war.[56] At least 62 people died in the resulting clashes between pro-government and opposition militias.[57]

On 21 May 2008, after five days of negotiation under Arab League mediation in Qatar, all major parties signed the Doha Agreement, which ended the fighting.[53] [57] Under the accord, both sides agreed to elect former army head

Michel Suleiman president and establish a national unity government with a veto share for the opposition.[53] This ended 18 months of political paralysis.[56] The agreement was a victory for opposition forces, who received concessions regarding the composition of the cabinet, Hezbollah's telecommunications network, and the airport security chief, increasing their political clout.[57]

2011 government collapse

In early January 2011, the national unity government collapsed after all ten opposition ministers and one presidential appointee resigned due to tensions stemming from the Special Tribunal for Lebanon, which was expected to indict Hezbollah members in the assassination of former prime minister Rafic Hariri.[58] The collapse plunged Lebanon into its worst political crisis since the 2008 fighting, and indicated further political gains for the Hezbollah-led opposition March 8 Alliance, which gained a parliamentary majority. The parliament elected Najib Mikati, the 8 March candidate, Prime Minister of Lebanon, making him responsible for forming a new government.[59]

Geography and climate

Lebanon is located in Western Asia, between latitudes 33° and 35° N, and longitudes 35° and 37° E. It is bordered by the Mediterranean Sea to the west along a 225-kilometre (140 mi) coastline, by Syria to the east and north, and by Israel to the south. The Lebanon-Syria border stretches for 375 kilometres (233 mi) and the Lebanon-Israel border for 79 kilometres (49 mi). The border with the Israeli-occupied Golan Heights is disputed by Lebanon in a small area called Shebaa Farms.[60]

Most of Lebanon's area is mountainous terrain,[61] except for the narrow coastline and the Beqaa Valley, which plays an integral role in Lebanon's agriculture. However, climate change and political differences threaten conflict over water resources in the Valley.[62]

Lebanon from space. Snow cover can be seen on the western and eastern mountain ranges

Lebanon has a moderate Mediterranean climate. In coastal areas, winters are generally cool and rainy whilst summers are hot and humid. In more elevated areas, temperatures usually drop below freezing during the winter with frequent, sometimes heavy snow; summers are warm and dry.[63] Although most of Lebanon receives a relatively large amount of rainfall annually (compared to its arid surroundings), certain areas in north-eastern Lebanon receive little because of the high peaks of the western mountain front blocking much of the rain clouds that originate over the Mediterranean Sea.[64]

Mountain scenery in Barouk

In ancient times, Lebanon housed large forests of the Cedars of Lebanon, which now serve as the country's national emblem.[65] In 2010, the Agriculture Ministry set a 10-year plan to increase the national forest coverage by 20% that is equivalent to the planting of two million new trees each year.[66] The plan, which was funded the U.S. development agency, USAID, and overseen by the U.S. Forest Service, and the Lebanon Reforestation Initiative, was inaugurated in 2011 by planting of seedlings, such as cedar, pine, wild almond, juniper, fir and oak, in five regions around Lebanon.[66]

A view from Beaufort Castle in south Lebanon

Late Cretaceous fish fossils beds of Lebanon are world famous, and are in the top twenty or thirty such location around the world.[67]

Government and politics

Lebanon is a parliamentary democracy, which implements a special system known as confessionalism.[68] This system is intended to deter sectarian conflict and attempts to fairly represent the demographic distribution of the 18 recognized religious groups in government.[69] [70] High-ranking offices are reserved for members of specific religious groups. The President, for example, has to be a Maronite Christian, the Prime Minister a Sunni Muslim, the Speaker of the Parliament a Shi'a Muslim, the Deputy Prime Minister and the Deputy Speaker of Parliament Greek Orthodox.[71] [72]

Lebanon's national legislature is the unicameral Parliament of Lebanon. Its 128 seats are divided equally between Christians and Muslims, proportionately between the 18 different denominations and proportionately between its 26 regions.[73] Prior to 1990, the ratio stood at 6:5 in favor of Christians; however, the Taif Accord, which put an end to the 1975–1990 civil war, adjusted the ratio to grant equal representation to followers of the two religions.[71] The Parliament is elected for a four-year term by popular vote on the basis of sectarian proportional representation.[5]

The Lebanese parliament building at the Place de l'Étoile.

The Grand Serail, the government headquarters in downtown Beirut.

The executive branch consists of the President, the head of state, and the Prime Minister, the head of government. The parliament elects the president for a non-renewable six-year term by a two-third majority. The president appoints the Prime Minister,[74] following consultations with the parliament. The President and the Prime Minister form the Cabinet, which must also adhere to the sectarian distribution set out by confessionalism.

On 27 June 2009, Lebanon's president Michel Suleiman appointed parliamentary majority leader Saad Hariri as prime minister after his pro-Western coalition, the March 14 Alliance, defeated a Hezbollah-led alliance in a June 2009 election.[75] In November, after five months of cabinet negotiations, Hariri formed a national unity government.[76] In January 2011, the government collapsed after all ten opposition ministers and one presidential appointee resigned due to tensions stemming from the Special Tribunal for Lebanon, which was expected to indict Hezbollah members in the assassination of former prime minister Rafic Hariri.[58]

Lebanon's judicial system is a mixture of Ottoman law, Napoleonic code, canon law and civil law. The Lebanese court system consists of three levels: courts of first instance, courts of appeal, and the court of cassation. The Constitutional Council rules on constitutionality of laws and electoral frauds. There also is a system of religious courts having jurisdiction over personal status matters within their own communities, with rules on matters such as marriage and inheritance.[77]

Foreign relations

Lebanon concluded negotiations on an association agreement with the European Union in late 2001, and both sides initialed the accord in January 2002. Lebanon also has bilateral trade agreements with several Arab states and is working toward accession to the World Trade Organization.

Lebanon enjoys good relations with virtually all of the other Arab countries (despite historic tensions with Libya, the Palestinians, Syria and Iraq), and hosted an Arab League Summit in March 2002 for the first time in more than 35 years. Lebanon is a member of the Francophone countries and hosted the Francophone Summit in October 2002 as well as the Jeux de la Francophonie in 2009.

Military

The Lebanese Armed Forces (LAF) has 72,100 active personnel,[78] including 1,100 in the air force, and 1,000 in the navy.[79] The Lebanese Armed Forces' primary missions include defending Lebanon and its citizens against external aggression, maintaining internal stability and security, confronting threats against the country's vital interests, engaging in social development activities, and undertaking relief operations in coordination with public and humanitarian institutions.[80]

Lebanon is a major recipient of foreign military aid.[81] With $400 million since 2005, it is the second largest per capita recipient of American military aid behind Israel.[82]

Governorates and districts

Lebanon is divided into six governorates (*mohaafazaat*, Arabic: محافظات —;singular *mohafazah*, Arabic: محافظة) which are further subdivided into twenty-five districts (*aqdya*—singular: *qadaa*).[83] The districts themselves are also divided into several municipalities, each enclosing a group of cities or villages. The governorates and their respective districts are listed below:

North

Akkar

Miniyeh-
Danniyeh

Zgharta

Koura

Tripoli

Bsharri

Batroun

**Mount
Lebanon**

Jbeil

Kesrwan

Matn

Beirut

♦

Baabda

Aley

Chouf

South

Jezzine

Sidon

Tyre

Beqaa

Hermel

Baalbek

Zahle

Western
Beqaa

Rashaya

Nabatieh

Hasbaya

Nabatieh

Marjeyoun

Bint
Jbeil

- Beirut Governorate
    - The Beirut Governorate is not divided into districts and is limited to the city of Beirut
- Nabatieh Governorate (*Jabal Amel*)
    - Bint Jbeil

- Hasbaya
- Marjeyoun
- Nabatieh
- Beqaa Governorate

  - Baalbek
  - Hermel
  - Rashaya
  - Western Beqaa (*al-Beqaa al-Gharbi*)
  - Zahle
- North Governorate (*al-Shamal*)

  - Akkar
  - Batroun
  - Bsharri
  - Koura
  - Miniyeh-Danniyeh
  - Tripoli
  - Zgharta
- Mount Lebanon Governorate (*Jabal Lubnan*)

  - Aley
  - Baabda
  - Byblos (*Jbeil*)
  - Chouf
  - Keserwan
  - Matn
- South Governorate (*al-Janoub*)

  - Jezzine
  - Sidon (*Saida*)
  - Tyre (*Sur*)

Economy

Economy of Lebanon
Tourism
Agriculture
Beirut Stock Exchange
Companies listed on BSE
Companies
Banque du Liban
Shipping

Topics of Lebanon
Culture - Geography
History - Politics

The urban population in Lebanon is noted for its commercial enterprise.[84] Over the course of time, emigration has yielded Lebanese "commercial networks" throughout the world.[85] As a result, remittances from Lebanese abroad to family members within the country total $8.2 billion[86] and account for one fifth of the country's economy.[87] Lebanon has the largest proportion of skilled labor among Arab States.[88]

Although Lebanon is ideally suited for agricultural activities in terms of water availability and soil fertility, as it possesses the highest proportion of cultivable land in the Arabic speaking world,[89] it does not have a large agricultural sector. Attracting only 12% of the total workforce,[90] agriculture is the least popular economic sector in Lebanon. It contributes approximately 11.7% of the country's GDP, also placing it in the lowest rank compared to other economic sectors. Major produce includes apples, peaches, oranges, and lemons.[9]

Industry in Lebanon is mainly limited to small businesses that reassemble and package imported parts. In 2004, industry ranked second in workforce, with 26% of the Lebanese working population,[90] and second in GDP contribution, with 21% of Lebanon's GDP.[9]

A combination of beautiful climate, many historic landmarks and World Heritage Sites continues to attract large numbers of tourists to Lebanon. In addition, Lebanon's strict financial secrecy and capitalist economy have given it significant, though no longer dominant, economic status among Arab countries. The thriving tourism and banking activities have naturally made the services sector the most important pillar of the Lebanese economy. The majority of the Lebanese workforce (nearly 65%)[90] attains employment in the services sector as a result of the abundant job opportunities. The GDP contribution, accordingly, amounts to roughly 67.3% of the annual Lebanese GDP.[9] However, dependence on the tourism and banking sectors leaves the economy vulnerable to political instability.[91]

The Kadisha Valley is a World Heritage Site

The 1975–1990 civil war heavily damaged Lebanon's economic infrastructure, cut national output by half, and all but ended Lebanon's position as a West Asian entrepôt and banking hub.[5] The subsequent period of relative peace enabled the central government to restore control in Beirut, begin collecting taxes, and regain access to key port and government facilities. Economic recovery has been helped by a financially sound banking system and resilient small- and medium-scale manufacturers, with family remittances, banking services, manufactured and farm exports, and international aid as the main sources of foreign exchange.[92]

Until the 2006 Lebanon War, Lebanon's economy witnessed excellent growth, with bank assets reaching over 75 billion US dollars,[93] By the end of the first half of 2006, the influx of tourists to Lebanon had already registered a 49.3% increase over 2005 figures (which was a low figure, making the 49.3% increase seem more spectacular than it was).[93] Market capitalization was also at an all time high, estimated at $10.9 billion at the end of the second quarter of 2006, just weeks before the fighting started.[93]

The war severely damaged Lebanon's fragile economy, especially the tourism sector. According to a preliminary report published by the Lebanese Ministry of Finance on 30 August 2006, a major economic decline was expected as a result of the fighting.[94]

Rafiq Hariri International Airport re-opened in September 2006, and the efforts to revive the Lebanese economy have proceeded at a slow pace. Major contributors to the reconstruction of Lebanon include Saudi Arabia (with US$ 1.5 billion pledged),[95] the European Union (with about $1 billion)[96] and a few other Persian Gulf countries with contributions of up to $800 million.[97]

According to the CIA World Factbook, Lebanon's 2010 public debt exceeded 150.7% of GDP, ranking fourth highest in the world as a percentage of GDP, though down from 154.8% in 2009.[5] Finance minister Mohammad Chatah stated that the debt reached $47 billion in 2008 and would increase to $49 billion if privatization of two telecoms companies did not occur.[98] The *Daily Star* wrote that exorbitant debt levels have "slowed down the economy and reduced the government's spending on essential development projects."[99]

Given the frequent security turmoil it has faced, the Lebanese banking system has adopted a conservative approach, with strict regulations imposed by the central bank to protect the economy from political instability. These regulations have generally left Lebanese banks unscathed by the Financial crisis of 2007–2010. Lebanese banks remain, under the current circumstances, high on liquidity and reputed for their security.[100] Consequently, Lebanon is one of the only seven countries in the world in which the value of the stock markets increased in 2008.[101] Moreover, in 2009, Lebanon hosted the largest number of tourists to date, eclipsing the previous record set before the Lebanese Civil War.[102] The Lebanese economy grew 8.5% in 2008 and a revised 9% in 2009[103] despite a global recession.[104] Furthermore, the World Bank estimated GDP growth in 2010 at 7%.[104] As of 31 August 2010, The Daily Star reported that The Economist Intelligence Unit (EIU) has released an updated outlook on the Lebanese economy, predicting real gross domestic produce (GDP) growth would reach 6.8% in 2010 and 5.8% in 2011.[105]

Oil has recently been discovered inland and in the seabed between Lebanon, Cyprus, Israel and Egypt and talks are underway between Cyprus and Egypt to reach an agreement regarding the exploration of these resources.The seabed separating Lebanon and Cyprus is believed to hold significant quantities of crude oil and natural gas.[106]

To boost the economy and increase foreign direct investments, the Lebanese government has established a national investment promotion agency, IDAL, the Investment Development Authority of Lebanon in 1994. It was established with the aim of promoting Lebanon as a key investment destination, and attracting facilitating, and retaining investments in the country. In 2001, Investment Law No.360[107] was enacted to reinforce the organisation's mission, providing a framework for regulating investment activities in Lebanon, and providing local and foreign investors alike with a range of incentives and business support services. In addition to its role as an investment promotion agency, IDAL was entrusted with the active promotion and marketing of Lebanese exports including but not limited to agricultural and agro-industrial products. IDAL enjoys financial and administrative autonomy and reports to the President of the Council of Ministers who exercises a tutorial authority over it.

Tourism

Between 2005 and 2007, Lebanon was in a state of political turmoil, resulting in a sharp fall in tourism. Over the course of 2008 Lebanon rebuilt its infrastructure mainly in the real estate and tourism sectors resulting in a comparatively robust post war economy. In 2009, the New York Times ranked Beirut the No. 1 travel destination worldwide Due to its Unique Nightlife and Hospitality.[108] In January 2010, the Ministry of Tourism announced that 1,851,081 tourists had visited Lebanon in 2009, a 39% increase from 2008, with Hotels reporting an occupancy rate of 95% in 2009. In March 2010, the Lebanon Opportunities review reported that 500,000 tourists had already come to Lebanon since the beginning of the year. Overall, Lebanon has seen

Baalbek Temple of Jupiter

an annual increase in tourism since 2006. The Ministry of Tourism said that more than 2.5 million tourists from the Gulf, Europe, Asia, Africa, and America visited in 2010.

Recently, after the long years of the civil war and reoccurring periods of political unrest in Lebanon, Lebanon has become an increasingly popular destination for tourism. Its rich history, historic sites, mild climate, along with other factors, have all made Lebanon currently one of the most visited countries in the Middle East. Lebanon, even in its post-war state, has managed to attract around 1,333,000 tourists in 2008, thus placing it as rank 79 out of 191 participating countries.[109] Statistics have shown that Lebanon's tourist attraction rate has been increasing rapidly and the Ministry of Tourism predicts that this ongoing trend will amplify in the coming years. Saudi Arabia and Jordan are the two most popular origin countries of foreign tourists to Lebanon.[110]

Sunset in Raouche

Education

Schools

All Lebanese schools are required to follow a prescribed curriculum designed by the Ministry of Education. Some of the 1400 private schools offer IB programs,[111] and may also add more courses to their curriculum with approval from the Ministry of Education. The main subjects taught are mathematics, sciences, Arabic, and at least one secondary language (either French or English).

The government introduces a mild form of selectivity into the curriculum by giving 11th graders choice between two "concentrations": sciences, humanities, and 12th graders choose between four concentrations: life sciences, general sciences, sociology and economics, and humanities and literature. The choices in concentration do not include major changes in the number of subjects taken (if at all). However, subjects that fall out of the concentration are given less weight in grading and are less rigorous, while subjects that fall within the concentration are more challenging and contribute significantly to the final grade.

Students go through three academic phases:

Name	Number of years	Annotations
Elementary	6	
Intermediate	3	students earn Intermediate Certification (Lebanese *Brevet*) at completion
Secondary	3	students who pass official exams earn a Baccalaureate Certificate (*Baccalauréat Libanais*) in the concentration they chose in 12th grade. Students studying at French-system schools or American-system schools may also graduate with a French Baccalaureate that is considered equivalent to the Lebanese Baccalaureate. Students can also graduate with an International Baccalaureate (current in some of the private schools).

The first eight years are, by law, compulsory.[9] Nevertheless, this requirement currently falls short of being fully enforced.

Higher education

Following secondary school, Lebanese students may choose to study at a university, a college, or a vocational training institute. The number of years to complete each program varies. While the Lebanese educational system offers a very high quality and international class of education, the local employment market lacks sufficient opportunities, thus encouraging many of the young educated to travel abroad.

Lebanon has forty-one nationally accredited universities, several of which are internationally recognized.[112] [113] The American University of Beirut (AUB) and the Université Saint-Joseph (USJ) were the first Anglophone and the first Francophone universities to open in Lebanon, respectively.[114] [115] Another prestigious and internationally recognized university is the Lebanese American University . The Lebanese American University is composed of two campuses, one in Beirut and the other in Byblos. Universities in Lebanon, both public and private, largely operate in French or English.[116]

The American University of Beirut is one of the highest-ranked and oldest universities in the Middle East. In 1862 American missionaries in Lebanon and Syria, under the American Board of Commissioners for Foreign Missions, asked Dr. Daniel Bliss to establish a college of higher learning that would include medical training. On 24 April 1863, while Dr. Daniel Bliss was raising money for the new college in the United States and England, the State of New York granted a charter for the Syrian Protestant College. The college, which was renamed the American University of Beirut in 1920, opened with a class of 16 students on 3 December 1866. Dr. Bliss served as its first president, from 1866 until 1902. The American University of Beirut (AUB) has been accredited as an institution since 2004 by the Commission on Higher Education of the Middle States Association of Colleges and Schools (3624 Market Street, Philadelphia, PA 19104, Tel. 267-284-5000). AUB's accreditation was most recently reaffirmed in June 2009, after the completion of an extensive self-study that was reviewed by educational experts chosen in consultation with Middle States. The University's next full accreditation cycle is due in 2018–19. Over the last several years, a number of University programs and faculties have also sought accreditation with more specialized bodies. The Faculty of Health Sciences' Graduate Public Health program became the first such program to be accredited by the Council on Education for Public Health (CEPH) outside of North America. Similarly, the Rafic Hariri School of Nursing became the first nursing school beyond American territories to have BSN and MSN programs accredited by the Commission on Collegiate Nursing Education (CCNE). In April 2009, undergraduate and graduate programs at the Suliman S. Olayan School of Business were accredited by the Association to Advance Collegiate Schools of Business (AACSB). Most recently, in July 2010, four undergraduate Engineering programs at AUB's Faculty of Engineering and Architecture were accredited by ABET Inc. (Accreditation Board for Engineering and Technology). In general, the accreditation process is intended to strengthen and sustain the quality and integrity of a university, faculty, or program, confirming that it is worthy of public confidence. AUB has been registered with and recognized by the New York State Education Department (NYSED) since 1863. Its degrees are recognized by the Lebanese government through the equivalence committees of the Ministry of Education and Higher Education.

At the English universities, students who have graduated from an American-style high school program enter at the freshman level to earn their baccalaureate equivalence from the Lebanese Ministry of Higher Education. This qualifies them to continue studying at the higher levels. Such students are required to have already taken the SAT I and the SAT II upon applying to college, in lieu of the official exams. On the other hand, students who have graduated from a school that follows the Lebanese educational system are directly admitted to the sophomore year. These students are still required to take the SAT I, but not the SAT II. The university academic degrees for the first stage are the Bachelor or the Licence, for the second stage are the Master or the DEA and the third stage is the doctorate.

The United Nations assigned Lebanon an education index of 0.871 in 2008. The index, which is determined by the adult literacy rate and the combined primary, secondary, and tertiary gross enrollment ratio, ranked the country 88th out of the 177 countries participating.[117]

Demographics

Identifying all Lebanese as ethnically Arab is a widely employed example of panethnicity since in reality, the Lebanese "are descended from many different peoples who have occupied, invaded, or settled this corner of the world," making Lebanon, "a mosaic of closely interrelated cultures".[118] While at first glance, this ethnic, linguistic, religious and denominational diversity might seem to cause civil and political unrest, "for much of Lebanon's history this multitudinous diversity of religious communities has coexisted with little conflict".[118]

The population of Lebanon was estimated to be 4,125,247 in July 2010,[5] however no official census has been taken since 1932 due to the sensitive confessional political balance between Lebanon's various religious groups.[119]

Population projection

Lebanese fertility declined from 4.23 in 1978 to within decimal points of the 2.1 children per woman level in 2000, and this was because "most of the female population [fell] into the better-educated groups", making Lebanon's fertility rate the lowest in the Arabic-speaking world.[120]

US Census Bureau, 2010 est.[121] :

- **2020**: 4,459,000
- **2030**: 4,512,000
- **2040**: 4,498,000
- **2050**: 4,389,000

United Nations, 2010 est.[122] :

- **2020**: 4,617,000
- **2030**: 4,713,000
- **2040**: 4,655,000
- **2050**: 4,414,000
- **2060**: 4,211,000
- **2070**: 4,113,000
- **2080**: 4,090,000
- **2090**: 3,989,000
- **2100**: 3,870,000

Religion

Lebanon's population is estimated to be 59.7% Muslim: 27% Sunni 27% Shia 5.7% Other (Shia, Sunni, Isma'ilite, Alawite, or Nusayri and non—Muslims with similar beliefs to the Muslim such as Druze), 39% Christian: (Maronite, Greek Orthodox, Melkite Catholic, Armenian Orthodox, Syriac Catholic, Armenian Catholic, Syriac Orthodox, Roman Catholic, Chaldean, Assyrian, Copt, Protestant), and 1.3% other.[5] Over the past 60 years, there has been a steady decline in the number of Christians as compared to Muslims, due to higher emigration rates among Christians, and a higher birth rate among the Muslim population.[124] The most recent study conducted by Statistics Lebanon, a Beirut-based research firm, found that approximately 27% of the population was Sunni, 27% Shi'a, 21% Maronite, 8% Greek Orthodox, 5% Druze, 5% Greek Catholic, and 7% other Christian sects.[124] There are 18 state-recognized religious sects – 4 Muslim, 12 Christian, 1 Druze, and 1 Jewish.[124]

Lebanon has the most religiously diverse society in the Middle East.[123]

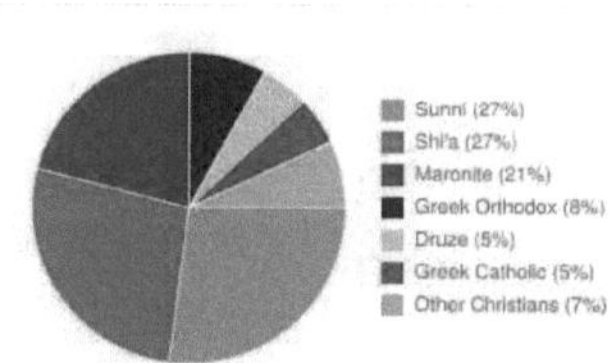

Religions in Lebanon by sect (2010).[124]

The Shi'a community is estimated to be 27%[124] of Lebanon's total population, and is often described as being the largest of Lebanon's Muslim communities,[125] or the largest of the 18 recognized religious sects in Lebanon.[126] Shi'a residents primarily live in South Beirut, the Beqaa Valley, and southern Lebanon.[125]

The Sunni community is estimated to be 27% of Lebanon's total population.[124] Sunni residents primarily live in West Beirut, the southern coast of Lebanon, and northern Lebanon.[125]

The Maronite community is estimated to be approximately 21% of Lebanon's total population.[124] Maronite residents tend to live in East Beirut and the mountains of Lebanon.[125] They are the largest Christian community in Lebanon.[125]

The Greek orthodox community is estimated to be approximately 8% of Lebanon's total population. Greek orthodox residents primarily live in Koura, Beirut, Zahleh, Rachaya, Matn, Aley, Akkar, Tripoli, Hasbaya and Marjeyoun. They are the second largest Christian community in Lebanon and the 4th largest religious community in the country.

Language

Article 11 of Lebanon's Constitution states that "Arabic is the official national language. A law determines the cases in which the French language may be used".[127] The majority of Lebanese people speak Lebanese Arabic, while formal Arabic is mostly used in magazines, newspapers, and formal broadcast media. Almost 40% of Lebanese are considered francophone, and another 15% "partial francophone," and 70% of Lebanon's secondary school use French as a second language of instruction.[128] By comparison, English is used as a secondary language in 30% of Lebanon's secondary schools.[128] The use of French is a legacy of the post-World War I League of Nations mandate over Lebanon given to France; as of 2004, some 20% of the population used French on a daily basis.[61]

English is increasingly used in science and business interactions, but French is still the language generally used by intellectuals.[129] Lebanese people of Armenian, Assyrian, or Greek descent often speak Armenian, Neo-Aramaic, or Greek with varying degrees of fluency. There are currently around 150,000 Armenians in Lebanon, or around 5% of the population.[130]

Diaspora

Millions of people of Lebanese descent are spread throughout the world, mostly Christians,[131] especially in Latin America.[132] Brazil has the largest expatriate population.[133] See Lebanese Brazilian. Large numbers of Lebanese migrated to West Africa, particularly in the Ivory Coast (home to over 100,000 Lebanese)[134] and Senegal (roughly 30,000 Lebanese).[135] Australia is home to over 270,000 Lebanese (1999 est.).[136]

Refugees

As of 2007, Lebanon was host to over 375,000 refugees and asylum seekers: 270,800 Palestinians, 50,000 from Iraq,[137] and 4,500 from Sudan. Lebanon forcibly repatriated more than 300 refugees and asylum seekers in 2007.[138]

In the last three decades, lengthy and destructive armed conflicts have ravaged the country. The majority of Lebanese have been affected by armed conflict; those with direct personal experience include 75% of the population, and most others report suffering a range of hardships. In total, almost the entire population (96%) has been affected in some way – either personally or because of the wider consequences of armed conflict.[139]

Culture

Overview

The area including modern Lebanon has been home to various civilizations and cultures for thousands of years. Originally home to the Phoenicians, and then subsequently conquered and occupied by the Assyrians, the Persians, the Greeks, the Romans, the Arabs, the Crusaders, the Ottoman Turks and most recently the French, Lebanese culture has over the millennia evolved by borrowing from all of these groups. Lebanon's diverse population, composed of different ethnic and religious groups, has further contributed to the country's festivals, musical styles and literature as well as cuisine. When compared to the rest of the Southwest Asia, Lebanese society as a whole is well educated and 91%[140] of the population was literate. Despite the ethnic, linguistic, religious and denominational diversity of the Lebanese, they "share an almost common culture. . . .". Lebanese Arabic is universally spoken while food, music, and literature are deep-rooted "in wider Mediterranean and Levantine norms. . . .".[118] Lebanese society is very modern and similar to certain cultures of Mediterranean Europe as the country is "linked ideologically and culturally to Europe through France, and its uniquely diverse ethnic and religious composition [create] a rare environment that [is] at once Arab and European.[141] It is often considered as Europe's gateway to Western Asia as well as Asia's gateway to the Western World.[142]

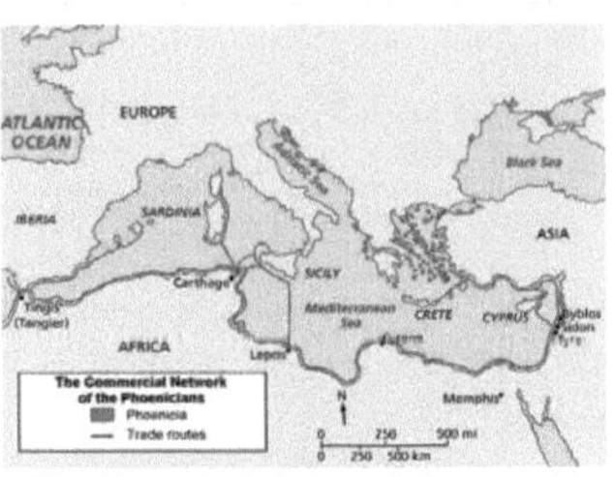

Phoenicia and its colonies

The original design

Arts and literature

By the turn of the 20th century, Beirut was vying with Cairo to be the major center for modern Arab thought, with many newspapers, magazines and literary societies. Additionally, Beirut became a thriving epicenter of Armenian culture with varied productions[143] that was exported to the Armenian diaspora.

In literature, Khalil Gibran, who was born in Bsharri, is particularly known for his book *The Prophet*, which has been translated into more than twenty different languages.[144] Several contemporary Lebanese writers have also achieved international success; including Elias Khoury, Amin Maalouf, Hanan al-Shaykh, and Georges Schehadé.

In art, Moustafa Farroukh was one of Lebanon's most prominent painters of the 20th century. Formally trained in Rome and Paris, he exhibited in venues from Paris to New York to Beirut over his career.

Khalil Gibran (April 1913)

Many more interesting and contemporary artists are currently active, such as Walid Raad a contemporary media artist currently residing in New York.

Two contemporary art exhibition centers, the Beirut Art Center (located in an industrial building painted in white near the Beirut river) and the Beirut Exhibition center (a very modern glass structure) in the BIEL area reflect the vibrant Lebanese contemporary art scene. These two centers are intended to host exhibitions and are a must in the world of international as well as local contemporary art.

Many art galleries also testify to the liveliness of the local art scene, exhibiting the works of new and talented artists such as Ayman Baalbaki, Akram Zaatari, Marwan Sahmarani, Nadim Asfar, Lamia Joreige, Jean Marc Nahas and many others.

These galleries are run by passionate gallerists such as Saleh Barakat (Agial), Neyla Kettaneh Kunigk, Fadi Mogabgab, Galerie Janine Rubeiz or the resounding Ayyam gallery whose owner is a Syrian national, one of the promoters of artistic renewal in this neighboring country.

Located in Foch Street in the Solidere area, FFA Private Bank is home to many temporary exhibitions of contemporary local artists as well as to a permanent display of paintings by Lebanese artists (Sahmarani, Baalbaki, Hannibal Srouji...) or foreign artists such as Fabienne Arietti's "Nasdaq". At the entrance of the bank's building (typical of the architecture of the old Beirut with a futuristic interior design), visitors are greeted by a strange security guard, piece of work from "ultra-realistic" New York sculptor Marc Sijan.

A Jean Dubuffet's huge sculpture can also be seen when visiting the atrium of Bank Audi Plaza, located in a beautiful contemporary building designed by Kevin Dash. By Strolling through the streets of the city one can find some interesting works such as sculptures of Michel Basbous in the Bank of Lebanon street.

Another initiative is Ashkal alwan, a Lebanese association for plastic arts and a platform for the creation and exchange of artistic practices.

It was founded by Christine Tohme, Marwan Rechmaoui, Rania Tabbara, Mustapha Yamout and Leila Mroueh Initially, Ashkal Alwan promoted and introduced the work of artists who have been engaged in critical art practices within the context of post-war Lebanon.

The Home Works Forum is a multidisciplinary platform that takes place in Beirut, Lebanon about every other year. it has evolved into one of the most vibrant platforms for research and exchange on cultural practices in the region and beyond.

The main languages being taught in schools and universities are listed as: Arabic, French and English.

Festivals

Music festivals, often hosted at historical sites, are a customary element of Lebanese culture.[145] Among the most famous are Baalbeck International Festival, Byblos International Festival, Beiteddine International Festival, Broumana Festival, Batroun Festival, Dhour Chwer Festival and Tyr Festival.[145] [146] These festivals are promoted by Lebanon's Ministry of Tourism, Lebanon Hosts about 15 Concerts from International Performers Each Year Ranking Number one for Nightlife in the Middle east and 6th Worldwide.[147]

Beiteddine Palace, venue of the Beiteddine Festival

Holidays

Lebanon has Christian and Muslim holidays; national holidays are also observed.

National flag

The national flag of Lebanon, created shortly after independence in 1943,[148] consists of three horizontal bands; the top and bottom bands are red and of equivalent size, each consisting of 1/4 of the flag's surface, while the larger, middle band is white with a green cedar tree fixed at its center and consists of 1/2 of the flag's surface.[5] The cedar tree, an emblem of Lebanon, symbolizes survival,[149] the white band symbolises the eternal snow on its mountain peaks and the peace that Lebanon seeks. Red symbolizes the blood shed for independence. The top and bottom of the cedar touch the edge of both red bands.[150]

Music

Music is pervasive in Lebanese society.[151] While traditional folk music remains popular in Lebanon, modern music reconciling Western and traditional Arabic styles, pop, and fusion are rapidly advancing in popularity.[152] Radio stations feature a variety of music, including traditional Lebanese, classical Arabic, Armenian [153] and modern French, English, American, and Latin tunes.[154] Prominent traditional musicians include Fairuz, an icon during the civil war, Sabah, Wadih El Safi, Majida El Roumi, and Najwa Karam who built an international audience for the genre.[151] Marcel Khalife, a musician who blends classical Arab music with modern sounds, boasts immense[155] popularity for his politically charged lyrics.[151] [152] Distinguished pop artists include Nancy Ajram, Haifa Wehbe, The 4 Cats—an all-female group—, Fadl Shaker, Elissa and Mika.[151]

According to the World Intellectual Property Organization, Lebanon's music industry is growing and could attain leading status in the region.[156] Lebanese performers are celebrated throughout the Arab World,[157] and with the notable exception of Egypt enjoy increasing regional popularity.[156] Rising demand for Arabic music outside Western Asia has provided Lebanese artists with a small but significant global audience. However, widespread piracy continues to inhibit the music industry's growth.[156]

Sports

Both summer and winter sports thrive in Lebanon because of the unique geography. In autumn and spring, for example, it is possible to go skiing in the morning and swimming in the Mediterranean Sea in the afternoon. At the competitive level, basketball and football are among Lebanon's most popular sports. In recent years, Lebanon has hosted the AFC Asian Cup and the Pan Arab Games.

Lebanon has six ski resorts, with opportunities also available for cross-country skiing, snowshoeing, and snowmobiling. In the summer, skilifts can be used to access hiking trails, with views stretching as far as Cyprus to the west and Syria to the east on clear days. Canoeing, cycling, rafting, climbing, swimming, sailing and caving are among the other common leisure sports in Lebanon. Adventure and extreme sports are also possible throughout the country. The Beirut Marathon is held every fall, drawing top runners from Lebanon and abroad. Race day is promoted as a fun, family event, and it has become a tradition for many to participate in costumes or outlandish clothing.

But the most important of sports, and the most popular in Lebanon is basketball, as the Lebanese National Team prevailed to qualify for the FIBA World Championship 3 times in a row. Considered as one of the basketball power houses in Asia, Lebanon was able to defeat strong teams like Venezuela and shell-shock France in what was considered to be the upset of the tournament, throwing an amazing encounter proving to be one of the most competitive teams. In 2010 FIBA World Championship, Lebanon defeated Canada national men's basketball team but failed to qualify to the second round. Rony Seikali is considered to be the best Lebanese basketball player of all time. Dominant Basketball teams in Lebanon are Sporting Al Riyadi Beirut, who are the current Arab champions, Club Sagesse who were able to earn the Asian and Arab championships before, along with Champville SC, Al Mouttahed Tripoli, and Hoops Club,and Byblos.

Dance is also a popular activity in Lebanon that may fall under the category of 'sports'.

Lebanon hosted the 2009 Jeux de la Francophonie from 27 September to 6 October.

Prominent Lebanese bodybuilders include Samir Bannout, Mohammad Bannout and Ahmad Haidar.

Rugby league has enjoyed growth in Lebanon with a seven team domestic competition. An international team made up of domestic players recently played a two match tour in Dubai. The Lebanon national rugby league team took part in the 2009 European Cup. After narrowly failing to qualify for the final, the team defeated Ireland to finish 3rd in the tournament.

Hazem El Masri, who is the National Rugby League's all time highest points scorer, moved from Lebanon to Australia as a child and has represented Lebanon at international level, including playing at the 2000 Rugby League

World Cup

Theatre

Theatre has existed in Lebanon since the first musical plays of Maroun Naccache, which were written and performed in the mid-1800s and are considered the birth of modern Arab theatre.[158]

Media

Lebanon is not only a regional center of media production but also the most liberal and free in the Arab world.[159] According to Press freedom's Reporters Without Borders, "the media have more freedom in Lebanon than in any other Arab country".[160] Despite its small population and geographic size, Lebanon plays an influential role in the production of information in the Arab world and is "at the core of a regional media network with global implications".[161]

Film

Cinema of Lebanon, according to film critic and historian, Roy Armes, was the only other cinema in the Arabic-speaking region, beside Egypt's, that could amount to a national cinema.[162] Cinema in Lebanon has been in existence since the 1920s, and the country has produced over 500 films,[163] some of which are:

- *West Beirut* – by Ziad Doueiri, released in 1998, received the Prix François Chalais at the Directors' fortnight of the Cannes Film Festival (1998)
- *Mabrouk Again* – by Hany Tamba, released in 2000
- *The Kite*– by Randa Chahal, released in 2003, received many prestigious awards including the Silver Lion, Prix de la paix- Gillo Pontecorvo and Prix de la Lanterne Magique at the Venice Film Festival (2003)
- *After Shave* – by Hany Tamba, released in 2005, received the 2006 French César Award for best foreign short film
- *Bosta* – by Philippe Aractingi, released in 2005
- *Under the Bombs* – by Philippe Aractingi, released in 2006
- *Caramel* – starring and directed by Nadine Labaki, released in 2007
- *Where Do We Go Now?* – starring and directed by Nadine Labaki, released in 2011, received the Cadillac People's Choice Award at the Toronto Film Festival (2011)

Internet

Lebanon was one of the first countries in the Arabic-speaking world to introduce internet and Beirut's newspapers were the first in the region to provide readers with web versions of their newspapers. By 1996, three newspapers from Lebanon were online, *Al Anwar*, *Annahar*, and *Assafir*, and by 2000, more than 200 websites provided news out of Lebanon.[161]

Publishing

The history of publishing in Lebanon dates back to 1610 when the first printing press was established at the Convent of Saint Anthony of Qozhaya in the Kadisha Valley, making its first publication, *Qozhaya Psalter* -the Bible's book of psalms, which was in both Syriac and Arabic, the first publication in the Middle East.[164] One of the first Arabic-script, printing presses in the region was founded in 1734 at The Convent of St. John in Khinshara where it remained in operation until1899.[165] In the second half of the nineteenth century, Beirut had become not only a multi-religious, commercial center but also an intellectual one, especially after the establishment of two private, higher education institutes, the American University of Beirut in 1864 and the Saint Joseph University in 1875, and it was this period that marked the emergence of Beirut's prolific press.[166] Lebanese publishers and journalists, along with Syrians, also played a major role in establishing the Egyptian press in the nineteenth century.[167] After independence, Beirut emerged as the epicenter of publishing in the Arab world, characterized by free and liberal media and literary scenes.[168] In the 1940s, Beirut was home to 39 newspapers as well as 137 periodicals and journals that were published in three languages.[168] Beirut also hosted the first book fair in the Arab world in 1956. By the early sixties, there were close to a hundred publishers and more than 250 printing presses in Lebanon.[168] Armenian publications also flourished in Beirut with over 44 publications, including dailies and periodicals.[169] Authors from Syria, Palestine and elsewhere in the Arab world found refuge in Lebanon's free and liberal publishing industry.[168] Lebanon's press became a huge industry despite the country's small size and has remained a haven for Arabic publishing.[167] The establishment of modern printing presses and sophisticated book distribution channels made Beirut a regional publishing leader, and gave the Lebanese publishers a dominant role in Arab publishing.[170] Lebanon hosts annually two important regional publishing events, the Beirut Book Fair and the Beirut Francophone Book Fair.[171]

Television

Television was introduced in Lebanon in 1959, with the launch of two privately-owned stations, CLT and Télé Orient that merged in 1977 into Télé Liban.[172] Lebanon has ten national television channels, most channels in Lebanon are affiliated or supported by certain political parties or alliances.

Notes

[1] "The Lebanese Constitution" (http://presidency.gov.lb/English/LebaneseSystem/Documents/Lebanese Constitution.pdf) (PDF). Presidency of Lebanon. . Retrieved 20 August 2011.

[2] (PDF) *World Population Prospects, Table A.1* (http://www.un.org/esa/population/publications/wpp2008/wpp2008_text_tables.pdf). 2008 revision. United Nations Department of Economic and Social Affairs. 2009. p. 17. . Retrieved 22 September 2010.

[3] "Lebanon" (http://www.imf.org/external/pubs/ft/weo/2011/01/weodata/weorept.aspx?sy=2008&ey=2011&scsm=1&ssd=1& sort=country&ds=.&br=1&c=446&s=NGDPD,NGDPDPC,PPPGDP,PPPPC,LP&grp=0&a=&pr.x=36&pr.y=7). International Monetary Fund. . Retrieved 30 April 2011.

[4] "HDRO (Human Development Report Office United Nations Development Programme" (http://hdr.undp.org/en/media/ HDR_2011_EN_Tables.pdf). United Nations. 2011. . Retrieved 2 November 2011.

[5] "Lebanon" (https://www.cia.gov/library/publications/the-world-factbook/geos/le.html). *The World Factbook*. Central Intelligence Agency. 31 January 2011. . Retrieved 16 February 2011.

[6] *Republic of Lebanon* is the most common term used by Lebanese government agencies. The term *Lebanese Republic*, a literal translation of the official Arabic and French names that is not used in today's world. Arabic is the most common language spoken among the citizens of Lebanon.

[7] McGowen, Afaf Sabeh (1989). "Historical Setting" (http://hdl.loc.gov/loc.gdc/cntrystd.lb). In Collelo, Thomas. *Lebanon: A Country Study*. Area Handbook Series (3rd ed.). Washington, D.C.: The Division. OCLC 18907889. . Retrieved 24 July 2009.

[8] Dumper, Michael; Stanley, Bruce E.; Abu-Lughod, Janet L. (2006). *Cities of the Middle East and North Africa*. ABC-CLIO. p. 104. ISBN 1576079198. "Archaeological excavations at Byblos (Jbeil) indicate that the site has been continually inhabited since at least 5000 B.C."

[9] "Background Note: Lebanon" (http://www.state.gov/r/pa/ei/bgn/35833.htm). U.S. Department of State. 22 March 2010. . Retrieved 4 October 2010.

[10] Moubayed, Sami (5 September 2007). "Lebanon douses a terrorist fire" (http://www.atimes.com/atimes/Middle_East/II05Ak01.html). *Asia Times*. . Retrieved 27 October 2009.

[11] Anna Johnson (2006). "Lebanon: Tourism Depends on Stability" (http://www.chron.com/disp/story.mpl/ap/fn/4297143.html). . Retrieved 31 October 2006.

[12] "Lebanon: Country Profile" (http://www.acdi-cida.gc.ca/lebanon#3). Canadian International Development Agency. . Retrieved 2 December 2006.

[13] "Deconstructing Beirut's Reconstruction: 1990–2000" (http://www.csbe.org/saliba/essay1.htm). Center for the Study of the Built Environment. . Retrieved 31 October 2006.

[14] [IMF projects Lebanon real GDP IMF projects Lebanon real GDP growth in 2011 at 1.5 percent]. The Daily Star. 24 September 2011

[15] "UAE justice system ranked above US in annual index" (http://www.thenational.ae/news/uae-news/courts/ uae-justice-system-is-ranked-above-us). .

[16] Room, Adrian (2005). *Placenames of the World: Origins and Meanings of the Names for 6,600 Countries, Cities, Territories, Natural Features and Historic Sites* (2nd ed.). McFarland. pp. 214–216. ISBN 9780786422487.

[17] Metzger, Bruce M.; Coogan, Michael D. (2004). *The Oxford guide to people & places of the Bible*. Oxford University Press. p. 178. ISBN 0195176103.

[18] Bienkowski, Piotr; Millard, Alan Ralph (2000). *Dictionary of the ancient Near East*. University of Pennsylvania Press. p. 178. ISBN 9780812235579.

[19] Ross, Kelley L. "The Pronunciation of Ancient Egyptian" (http://www.friesian.com/egypt.htm). *The Proceedings of the Friesian School, Fourth Series*. Friesian School. . Retrieved 20 January 2009.

[20] Godefroy Zumoffen (1926). *Géologie du Liban* (http://books.google.com/books?id=4W-HQAAACAAJ). H. Barrère. . Retrieved 22 July 2011.

[21] Lorraine Copeland; P. Wescombe (1965). *Inventory of Stone-Age sites in Lebanon, p. 40* (http://books.google.com/books?id=6YsRRwAACAAJ). Imprimerie Catholique. . Retrieved 21 July 2011.

[22] "Archaeological Virtual Tours: Byblos" (http://web.archive.org/web/20080223164318/http://destinationlebanon.gov.lb/eng/Byblos/History.asp). Destinationlebanon.gov.lb. Archived from the original (http://destinationlebanon.gov.lb/eng/Byblos/History.asp) on 23 February 2008. . Retrieved 14 October 2008.

[23] About.com (1987). "Lebanon in Ancient Times" (http://ancienthistory.about.com/library/bl/bl_lebanonphoenicians.htm). Retrieved 17 December 2006.

[24] "Photographs in History" (Arabic) – sixth edition 1999. P. 76.

[25] Dreaming of Greater Syria (http://english.aljazeera.net/focus/arabunity/2008/02/2008525183842614205.html). Abdul-Ilah Saadi. *Al Jazeera English*.

[26] Chorbishop Seely Beggiani (2005). "Aspects of Maronite History (Part Eleven) The twentieth century in Western Asia" (http://www.stmaron.org/marhist11.html). Retrieved 24 January 2007.

[27] "Glossary: Cross-Channel invasion" (http://www.pbs.org/behindcloseddoors/glossary.html). Public Broadcasting Service. . Retrieved 17 October 2009.

[28] Harb, Imad (March 2006). "Lebanon's Confessionalism: Problems and Prospects" (http://web.archive.org/web/20080709034419/http://www.usip.org/pubs/usipeace_briefings/2006/0330_lebanon_confessionalism.html). *USIPeace Briefing*. United States Institute of Peace. Archived from the original (http://www.usip.org/publications/lebanons-confessionalism-problems-and-prospects) on 9 July 2008. . Retrieved 20 January 2009.

[29] "Background Note: Lebanon" (http://www.state.gov/r/pa/ei/bgn/35833.htm). *Bureau of Near Eastern Affairs*. U.S. Department of State. January 2009. . Retrieved 31 January 2010.

[30] Morris, p. 524

[31] Morris, p. 259

[32] Morris, p. 260

[33] "Lebanon Exiled and suffering: Palestinian refugees in Lebanon" (http://www.amnestyusa.org/document.php?lang=e&id=ENGMDE180102007) (in amn). *Amnesty International USA*. Amnesty International. 2007. . Retrieved 9 August 2009.

[34] "Lebanon: Exiled and suffering: Palestinian refugees in Lebanon" (http://www.amnestyusa.org/document.php?lang=e&id=ENGMDE180102007). Amnestyusa.org. . Retrieved 21 December 2010.

[35] al-Issawi, Omar (4 August 2009). "Lebanon's Palestinian refugees" (http://english.aljazeera.net/focus/2009/05/2009527115531294628.html). Al Jazeera. . Retrieved 21 August 2009.

[36] *Time* (1991). "After the War, the Mop-Up" (http://www.cedarland.org/time.html#9). Retrieved 30 November 2006.

[37] UN IRIN news. "Lebanon: Haven for foreign militants" (http://www.irinnews.org/Report.aspx?ReportId=72218). 17 May 2007.

[38] Council on Foreign Relations (2006). "The Future of Lebanon" (http://www.foreignaffairs.org/20061101faessay85602/paul-salem/the-future-of-lebanon.html). Retrieved 18 December 2006.

[39] Source: IMF – World Economic Outlook

[40] People's Daily (2000). "Lebanese Troops Patrol Near Fatma Gate Along Border With Israel" (http://english.peopledaily.com.cn/english/200008/10/eng20000810_47889.html). Retrieved 18 December 2006.

[41] Israel Ministry of Foreign Affairs (2000). "Withdrawal from Lebanon: Press Briefing by Foreign Minister David Levy" (http://www.mfa.gov.il/MFA/Government/Speeches+by+Israeli+leaders/2000/Withdrawal+from+Lebanon-+Press+Briefing+by+FM+Levy.htm). Retrieved 1 November 2006.

[42] The key to Shebaa (http://english.aljazeera.net/English/archive/archive?ArchiveId=11545), Al-Jazeera online. Retrieved 1 April 2007.

[43] Ross, Oakland (9 October 2007). "Language of murder makes itself understood" (http://www.thestar.com/article/264773). *Toronto Star*. . Retrieved 2 February 2009. "Like a wound that just won't heal, a large expanse patch of fresh asphalt still mottles the grey surface of Rue Minet el-Hosn, where the street veers west around St. George Bay. The patch marks the exact spot where a massive truck bomb exploded 14 February 2005, killing prime minister Rafik Hariri and 22 others and gouging a deep crater in the road."

[44] CBC News Indepth (2006). "Recent background on Syria's presence in Lebanon" (http://www.cbc.ca/news/background/lebanon/lebanon_syria.html). Retrieved 10 December 2006.

[45] this MEMRI bulletin (http://memri.org/bin/articles.cgi?Page=archives&Area=ia&ID=IA21005), includes several statements and sources.

[46] "Press Release SC/8353" (http://www.un.org/News/Press/docs/2005/sc8353.doc.htm) (Press release). United Nations – Security Council. 7 April 2005. . Retrieved 19 January 2009.

[47] Mehlis, Detlev (19 October 2005). "Report of the International Independent Investigation Commission established pursuant to Security Council resolution 1595" (http://web.archive.org/web/20080228173759/http://domino.un.org/UNISPAl.NSF/ fd807e46661e3689852570d00069e918/308be5d60f79289b852570a5005d0d00!OpenDocument). United Nations Information System on the Question of Palestine. Archived from the original (http://domino.un.org/UNISPAl.NSF/fd807e46661e3689852570d00069e918/ 308be5d60f79289b852570a5005d0d00!OpenDocument) on 28 February 2008. . Retrieved 2 February 2009. "It is the Commission's view that the assassination of 14 February 2005 was carried out by a group with an extensive organization and considerable resources and capabilities. [...] Building on the findings of the Commission and Lebanese investigations to date and on the basis of the material and documentary evidence collected, and the leads pursued until now, there is converging evidence pointing at both Lebanese and Syrian involvement in this terrorist act." "Report of the International Independent Investigation Commission established pursuant to Security Council resolution 1595" (http://www.un.org/News/dh/docs/mehlisreport/) (PDF). .

[48] "Syria begins Lebanon withdrawal" (http://news.bbc.co.uk/2/hi/middle_east/4342705.stm). BBC News. 12 March 2005. . Retrieved 11 December 2006.

[49] CNN (2005) Last Syrian troops leave Lebanon (http://web.archive.org/web/20080726023249/http://www.cnn.com/2005/WORLD/ meast/04/26/lebanon/). CNN. April 27, 2005. Retrieved 11 December 2006.

[50] 2005: Bassel Fleihan, Lebanese legislator and Minister of Economy and Commerce; Samir Kassir, Columnist and Democratic Left Movement leader; George Hawi, former head of Lebanese Communist Party; Gibran Tueni, Editor in Chief of "An Nahar" newspaper. 2006: Pierre Gemayel, Minister of Industry. 2007: Walid Eido, MP; Antoine Ghanim, MP.

[51] "Security Council calls for end to hostilities between Hizbollah, Israel" (http://www.un.org/News/Press/docs/2006/sc8808.doc.htm). UN — Security Council, Department of Public Information. 11 August 2006. . Retrieved 19 January 2009.

[52] UN IRIN news. "Life set to get harder for Nahr al-Bared refugees" (http://www.irinnews.org/Report.aspx?ReportId=81306). Retrieved 5 November 2008.

[53] Ruff, Abdul (1 June 2008). "Lebanon back to Normalcy?" (http://www.globalpolitician.com/24841-lebanon). *Global Politician*. . Retrieved 19 October 2009.

[54] "Beirut street clashes turn deadly" (http://www.france24.com/en/ 20080509-gunmen-force-shutdown-pro-government-tv-lebanon-unrest&navi=MONDE?q=node/1671710). *France 24*. . Retrieved 9 May 2008.

[55] Martínez, Beatriz; Francesco Volpicella (September 2008). "Walking the tight wire — Conversations on the May 2008 Lebanese crisis" (http://www.tni.org/article/walking-tight-wire). Transnational Institute. . Retrieved 9 May 2010.

[56] Abdallah, Hussein (22 May 2008). "Lebanese rivals set to elect president after historic accord" (http://www.dailystar.com.lb/article. asp?edition_id=1&categ_id=2&article_id=92308). *The Daily Star*. . Retrieved 19 October 2009.

[57] Worth, Robert; Nada Bakri (16 May 2008). "Feuding Political Camps in Lebanon Agree to Talk to End Impasse" (http://www.nytimes. com/2008/05/16/world/middleeast/16lebanon.html). *New York Times*. . Retrieved 19 October 2009.

[58] "Hezbollah and allies topple Lebanese unity government" (http://www.bbc.co.uk/news/world-middle-east-12170608). *BBC*. 12 January 2011. . Retrieved 12 January 2011.

[59] Bakri, Nada (12 January 2011). "Resignations Deepen Crisis for Lebanon" (http://www.nytimes.com/2011/01/13/world/middleeast/ 13lebanon.html?pagewanted=1&_r=1). *New York Times*. . Retrieved 12 January 2011.

[60] Telegraph (2000) "Israel's Withdrawal from Lebanon Given UN's Endorsement" (http://www.telegraph.co.uk/news/worldnews/ middleeast/lebanon/1343868/Israel's-withdrawal-from-Lebanon-given-UN's-endorsement.html). Retrieved 1 November 2006.

[61] "Lebanon" (http://www.britannica.com/EBchecked/topic/334152/Lebanon). *Encyclopædia Britannica*. 2011. .

[62] UN IRIN news. "Climate change and politics threaten water wars in Bekaa" (http://www.irinnews.org/Report.aspx?ReportId=82682). Retrieved 1 February 2009.

[63] (Bonechi et al.) (2004) *Golden Book Lebanon*, p. 3, Florence, Italy: Casa Editrice Bonechi. ISBN 88-476-1489-9

[64] Country Studies US. "Lebanon — Climate" (http://countrystudies.us/lebanon/31.htm). Retrieved 5 November 2006.

[65] Blue Planet Biomes. "Lebanon Cedar — Cedrus libani" (http://www.blueplanetbiomes.org/lebanon_cedar.htm). Retrieved 10 December 2006.

[66] Olivia Alabaster Lebanon begins landmark reforestation campaign (http://www.dailystar.com.lb/Article.aspx?id=155258). The Daily Star. November 26, 2011

[67] American University of Beirut "The geology of Lebanon a summary" (http://ddc.aub.edu.lb/projects/geology/geology-of-lebanon/). Retrieved 7 August 2010.

[68] Bureau of Democracy, Human Rights, and Labor (2002). "Country Reports on Human Rights Practices — 2002: Lebanon" (http://www. state.gov/g/drl/rls/hrrpt/2002/18281.htm). Retrieved 3 January 2007.

[69] Lijphart, Arend. *Consociational Democracy*, in "World Politics", Vol. 21, No. 2 (January 1969), pp. 207–225.

[70] Lijphart, Arend. *Multiethnic democracy*, in S. Lipset (ed.), "The Encyclopedia of Democracy". London, Routledge, 1995, Volume III, pp. 853–865.

[71] Imad Harb (March 2006). Lebanon's Confessionalism: Problems and Prospects (http://web.archive.org/web/20090322103601/http:// www.usip.org/pubs/usipeace_briefings/2006/0330_lebanon_confessionalism.html). United States Institute of Peace. Retrieved 3 January 2007.

[72] Marie-Joëlle Zahar. "Chapter 9 Power sharing in Lebanon: Foreign protectors, domestic peace, and democratic failure" (http://www.prio. no/files/file46602_zahar_-_power_sharing_in_lebanon.doc). (DOC) Retrieved 3 January 2007.

[73] Eager Lebanese race to polls to cast their ballots (http://www.alarabiya.net/save_pdf.php?cont_id=75167). AlArabbia. 7 June 2009. Retrieved 7 June 2009.

[74] UNDP. "Democratic Governance, Elections, Lebanon" (http://www.pogar.org/countries/theme.asp?th=3&cid=9). Retrieved 15 December 2008.

[75] "Pro-West prime minister to lead Lebanon" (http://www.msnbc.msn.com/id/31582576/ns/world_news-mideastn_africa/). *MSNBC*. 27 June 2009. . Retrieved 8 August 2009.

[76] "Lebanon's national-unity cabinet formed" (http://nowlebanon.com/NewsArticleDetails.aspx?ID=125341). *NOW Lebanon*. 9 November 2009. . Retrieved 10 November 2009.

[77] Carnegie Endowment for International Peace. *Arab Political Systems: Baseline Information and Reforms — Lebanon* (http://www.carnegieendowment.org/files/Lebanon_APS.doc). Carnegie Endowment for International Peace. . Retrieved 4 July 2009.

[78] "Lebanese Armed Forces, CSIS (Page 12)" (http://www.csis.org/media/csis/pubs/050323_memilbaldefine[1].pdf) (PDF). 21 October 2006. .

[79] Stinson, Jefferey (1 August 2006). "Lebanese forces may play bigger role in war" (http://www.usatoday.com/news/world/2006-08-01-lebanon-forces_x.htm). *USA TODAY*. . Retrieved 22 August 2009.

[80] "LAF Mission" (http://www.lebarmy.gov.lb/english/mission.asp). Lebanese Armed Forces. . Retrieved 19 May 2009.

[81] Lanteaume, Sylvie (4 August 2009). "US military aid at stake in Lebanon elections" (http://www.google.com/hostednews/afp/article/ALeqM5gfs5KFnHsffexKiQgh0f4eWb4Nbw). Agence France-Presse. . Retrieved 22 August 2009.

[82] Schenker, David (3 October 2008). "The Future of U.S. Military Aid to Lebanon" (http://www.washingtoninstitute.org/templateC05.php?CID=2933). *Washington Institute for Near East Policy*. . Retrieved 9 August 2009.

[83] USAID Lebanon. "USAID Lebanon—Definitions of Terms used" (http://lebanon.usaid.gov/(jakkco45gisweaychxvyoq55)/files/activities.aspx). Retrieved 17 December 2006.

[84] U.S. Department of State (1994) Header: People, 4th paragraph (http://dosfan.lib.uic.edu/ERC/bgnotes/nea/lebanon9401.html). Retrieved 3 December 2006.

[85] Background Note: Lebanon "www.washingtoninstitute.org" (http://www.washingtoninstitute.org/documents/41e1aa0d7d676.pdf). Retrieved 3 December 2006.

[86] International Organization for Migration. "Lebanon — Facts and Figures" (http://www.iom.int/jahia/Jahia/pid/426). Retrieved 13 June 2009.

[87] Reuters. "FACTBOX: Facts on Lebanon's economy" (http://www.reuters.com/article/wtMostRead/idUSTRE5570SJ20090608). Retrieved 13 June 2009.

[88] United Nations Population Fund. Lebanon — Overview (http://web.archive.org/20080207015156/http://www.unfpa.org/profile/lebanon.cfm?Section=1) at the Wayback Machine (*archived February 7, 2008*)

[89] Federal Research Division of the Library of Congress, U.S.A. 1986–1988. countrystudies.us (http://countrystudies.us/lebanon/71.htm). Retrieved 2 December 2006.

[90] Jean Hayek et al, 1999. The Structure, Properties, and Main Foundations of the Lebanese Economy. In *The Scientific Series in Geography, Grade 11*, 110–114. Beirut: Dar Habib.

[91] "Lebanon" (http://www.acdi-cida.gc.ca/lebanon) (Governmental). *Canadian International Development Agency*. Government of Canada. 28 May 2009. . Retrieved 24 August 2009.

[92] *CIA World Factbook 2001* (http://web.archive.org/web/20070604195718/http://teacherlink.ed.usu.edu/tlresources/reference/2001WorldFactbook/LEBANON.PDF). Retrieved 4 December 2006.

[93] Bank Audi (2006). "Lebanon Economic Report: 2nd quarter, 2006" (http://www.audi.com.lb/geteconomy/quarterly/lebanon.pdf). Retrieved 27 November 2005.

[94] Lebanese Ministry of Finance (2006). "Impact of the July Offensive on the Public Finances in 2006" (http://www.lebanonundersiege.gov.lb/documents/ImpactonfinanceReport-Englishversion-06.pdf). Retrieved 24 September 2006.

[95] Joseph S. Mayton Saudi Arabia Key Contributor To Lebanon's Reconstruction (http://web.archive.org/web/20070928180742/http://www.cynews.com/news/7005070415/). Cyprus News (2006). Retrieved 26 November 2006.

[96] Patrick Lannin Donors pledge more than $940 million for Lebanon (http://reliefweb.int/node/434440). Reuters. 31 August 2006. Retrieved 26 November 2006.

[97] Ain-Al-Yaqeen (2006). "The Custodian of the Two Holy Mosques Reviews with the Jordanian King the Situation in Lebanon..." (http://web.archive.org/web/20061020061315/http://www.ain-al-yaqeen.com/issues/20060825/feat2en.htm). Retrieved 27 November 2006.

[98] Bayoumy, Yara (2 January 2009). "RPT-UPDATE 1-Lebanon public debt at $47 bln end-2008-minister" (http://in.reuters.com/article/asiaCompanyAndMarkets/idINL217217120090102?pageNumber=1&virtualBrandChannel=0). Reuters. . Retrieved 18 October 2009.

[99] Daily Star Staff (20 May 2004). "IMF: Lebanon's debt alarming" (http://www.cggl.org/scripts/new.asp?id=227). *The Daily Star* (Center for Democracy and the Rule of Law). . Retrieved 18 October 2009.

[100] "Lebanon 'immune' to financial crisis" (http://news.bbc.co.uk/2/hi/middle_east/7764657.stm). BBC News. 5 December 2008. . Retrieved 28 January 2010.

[101] Cooper, Kathryn (5 October 2008). "Where on earth can you make a decent return?" (http://www.timesonline.co.uk/tol/money/investment/article4881201.ece). *The Sunday Times* (London). . Retrieved 28 January 2010.

[102] "Lebanon Says 2009 Was Best on Record for Tourism" (http://abcnews.go.com/Travel/wireStory?id=9601315). Associated Press. ABC News. 19 January 2010. . Retrieved 1 February 2010.

[103] Bayoumy, Yara (2 February 2010). "Lebanon central bank sees GDP growth topping 5 percent in 2010" (http://www.beirut-online.net/ portal/article.php?id=6619). *Reuters* (Beirut Online). . Retrieved 1 March 2010.

[104] Derhally, Massoud (21 January 2010). "Foreign Funds to Spur Growth in Lebanon, Salameh Says" (http://www.bloomberg.com/apps/ news?pid=20601104&sid=a3.ULyFK18Pk). *Bloomberg*. . Retrieved 1 February 2010.

[105] "Business Articles – EIU raises Lebanon's 2010 real GDP growth forecast" (http://www.dailystar.com.lb/article.asp?edition_id=1& categ_id=3&article_id=118808#axzz0yroZrvwP). The Daily Star. 31 August 2010. . Retrieved 21 December 2010.

[106] "The Next Big Lebanon-Israel Flare-Up: Gas" (http://www.time.com/time/world/article/0,8599,2061187,00.html). *Time*. 6 April 2011. .

[107] "Investment Law No.360" (http://www.idal.com.lb/OurProfile.aspx?ID=76). . Retrieved 29 July 2011.

[108] Zach Wise and Miki Meek/The New York Times (11 January 2009). "The 44 Places to Go in 2009 – Interactive Graphic" (http://www. nytimes.com/interactive/2009/01/11/travel/20090111_DESTINATIONS.html). *The New York Times*. . Retrieved 21 December 2010.

[109] "Tourist arrivals statistics – Countries Compared" (http://www nationmaster.com/graph/eco_tou_arr-economy-tourist-arrivals). NationMaster. . Retrieved 2011-11-04.

[110] "Business :: Lebanon :: Hospitality revenues plunge 40 percent in 2011" (http://www.dailystar.com.lb/Business/Lebanon/2011/Jul-16/ Hospitality-revenues-plunge-40-percent-in-2011.ashx#axzz1SE0e7i00). The Daily Star. 2011-07-16. . Retrieved 2011-11-04.

[111] Samidoun (2006). "Aid groups scramble to fix buildings; fill backpacks before school bell rings" (http://www.samidoun.org/?q=node/ 812). Retrieved 9 December 2006.

[112] Infopro Management. "Lebanon Opportunities – Business Information" (http://www.opportunities.com.lb/Lebanon/bhb/initdoc. asp?catId=21). Retrieved 30 January 2007.

[113] (Arabic)Lebanese Directory of Higher Education. Decrees (http://web.archive.org/web/20071211044854/http://www.higher-edu. gov.lb/Marasim.html). higher-edu.gov.lb. Retrieved 30 January 2007.

[114] eIFL.net Regional Workshop (2005). "Country Report: Lebanon" (http://www.eifl.net/docs/ collaborative_management_of_electronic_resources.ppt). Retrieved 14 December 2006.

[115] Université Saint-Joseph. 125 years of history – A timeline (http://web.archive.org/web/20060706105323/http://www.usj.edu.lb/ english/history.php). Retrieved 8 December 2006.

[116] "Yalla! Students" (http://yalla10.yalla.com.lb/students/abroad/english/lebanon.html). Retrieved 15 December 2006.

[117] "Human development indicators Lebanon" (http://hdrstats.undp.org/countries/country_fact_sheets/cty_fs_LBN.html). *United Nations Development Programme, Human Development Reports*. . Retrieved 17 November 2008.{ }

[118] Jamie Stokes (June 2009). *Encyclopedia of the Peoples of Africa and the Middle East: L to Z* (http://books.google.com/ books?id=stl97FdyRswC&pg=PA406). Infobase Publishing. pp. 406–. ISBN 978-0-8160-7158-6. . Retrieved 11 December 2011.

[119] Lebanon : Overview (http://www.unhcr.org/refworld/country,,,COUNTRYPROF,LBN,4562d8cf2,4954ce52c,0.html). Minority Rights Group International, *World Directory of Minorities and Indigenous Peoples*.

[120] Part Two – Population projections of countries and their coastal regions. Lebanon (http://www.planbleu.org/publications/demo_uk_lbn. pdf). planbleu.org. pp. 71–72

[121] "International Programs – U.S. Census Bureau" (http://www.census.gov/ipc/www/idb/informationGateway.php). Census.gov. . Retrieved 2011-11-04.

[122] (http://esa.un.org/unpd/wpp/unpp/p2k0data.asp)

[123] Dralonge, Richard N. (2008). *Economics and Geopolitics of the Middle East*. New York: Nova Science Publishers. p. 150. ISBN 1604560762. "Lebanon, with a population of 3.8 million, has the most religiously diverse society in the Middle East, comprising 17 recognized religious sects."

[124] Lebanon – International Religious Freedom Report 2010 (http://www.state.gov/g/drl/rls/irf/2010/148830.htm) U.S. Department of State. Retrieved on 14 February 2010.

[125] McGowen, Afaf Sabeh (1989). "Glossary" (http://hdl.loc.gov/loc.gdc/cntrystd.lb). In Collelo, Thomas. *Lebanon: A Country Study*. Area Handbook Series (3rd ed.). Washington, D.C.: The Division. OCLC 18907889. . Retrieved 30 September 2010.

[126] Winslow, Charles (1996). "Appendix I: Religious Sects of Lebanon" (http://books.google.com/books?id=MY4hzHa14KkC&lpg=PP1& pg=PA298). *Lebanon: War and Politics in a Fragmented Society*. Psychology Press. p. 298. ISBN 0415144035. . Retrieved 16 October 2010.

[127] "Article 11 of the Lebanese Constitution" servat.unibe.ch (http://www.servat.unibe.ch/icl/le00000_.html#A011_), Retrieved 28 June 2008.

[128] *The Story of French* (http://books.google.com/books?id=NN5oc0HFC7QC&pg=PA311). Macmillan. 2008. p. 311. ISBN 9780312341848. . Retrieved 14 December 2010.

[129] Nadeau, Jean-Benoît; Jean-Benoît Nadeau, Julie Barlow (2006). *Plus ça change* (http://books.google.com/?id=F_luCRxg6Q4C& pg=PA317). Robson. p. 483. ISBN 1861059175. . Retrieved 26 January 2010.

[130] Armenians jump Lebanon's divide (http://news.bbc.co.uk/2/hi/middle_east/8000507.stm). Not counted in the above-mentioned demographic survey of 2007 are the Armenian Christians and Syrian Orthodox who number 4% and 1% of the population. There are small numbers of Protestants and members of the Church of the East. *BBC News*. 16 April 2009.

[131] Van Dusenbery Senior Seminar: Transnational Migration and Diasporic Communities (http://web.archive.org/web/20090115011357/ http://www.hamline.edu/cla/academics/international_studies/diaspora2002/Lebanese/Paper.htm). Hamline University (2002-12-18)

[132] The world's successful diasporas (http://www.managementtoday.co.uk/news/648273/). *Management Today*. 3 April 2007.

[133] Marina Sarruf (2006). Estimated at about 15 million, the lebanese diaspora is composed mostly of christian, who started a mass emigration movement in 1850's. "Brazil Has More Lebanese than Lebanon" (http://web.archive.org/web/20061013002652/http://melbourne. indymedia.org/news/2006/07/117345.php). Retrieved 30 November 2006.

[134] Ivory Coast – The Levantine Community (http://countrystudies.us/ivory-coast/72.htm). Source: *U.S. Library of Congress.*

[135] Naomi Schwarz Lebanese Immigrants Boost West African Commerce (http://web.archive.org/web/20080323142904/http://www. voanews.com/english/archive/2007-07/2007-07-10-voa46.cfm?CFID=213232878&CFTOKEN=70264387). voanews.com. 10 July 2007

[136] Australian Population: Ethnic Origins (http://elecpress.monash.edu.au/pnp/free/pnpv7n4/v7n4_3price.pdf). (PDF)

[137] "Surviving in the city: A review of UNHCR's operation for Iraqi refugees in urban areas of Jordan, Lebanon and Syria" (http://www. unhcr.org/4a69ad639.pdf) (PDF). . Retrieved 21 December 2010.

[138] "World Refugee Survey 2008" (http://web.archive.org/web/20081221170955/http://www.refugees.org/article.aspx?id=2114& subm=179&area=Investigate). U.S. Committee for Refugees and Immigrants. 19 June 2008. .

[139] Lebanon, Opinion survey 2009 (http://www.icrc.org/Web/eng/siteeng0.nsf/htmlall/views-from-field-report-240609/$File/ Our-World-Views-from-Lebanon-I-ICRC.pdf), by ICRC and Ipsos

[140] Anthony J. Salim (1 June 2001). *Captivated by Your Teachings: A Resource Book for Adult Maronite Catholics* (http://books.google. com/books?id=7fS_2FisKoUC&pg=PA104). Paulist Press. pp. 104–. ISBN 978-1-893757-25-7. . Retrieved 11 December 2011.

[141] Davis, Craig S. *The Middle East For Dummies*

[142] Lebanon Culture. hangoverguide.com (http://hangoverguide.com/lebanon/), 18 December 2006.

[143] Migliorino, p. 166

[144] The Hindu (5 January 2003). Called by life (http://web.archive.org/web/20100812094242/http://www.hinduonnet.com/thehindu/lr/ 2003/01/05/stories/2003010500320500.htm). The Hindu. January 5, 2003. Retrieved 8 January 2007.

[145] Sheehan, Sean; Latif (30 August 2007). "Leisure" (http://books.google.com/?id=cA-RDzlwVVAC&pg=PA123). *Lebanon.* Cultures of the World. **13**. Zawiah. Marshall Cavendish Children's Books. p. 123. ISBN 9780761420811. .

[146] Carter, Terry; Dunston Lara (1 August 2004). "Getting Started" (http://books.google.com/?id=EskzgI-229IC&pg=PA11). *Syria & Lebanon.* Guidebook Series. Humphreys Andrew (2 ed.). Lonely Planet. p. 11. ISBN 9781864503333. .

[147] "Lebanon Summer & Winter Festivals" (http://www.lebanon-tourism.gov.lb/news/Details.aspx?NewsId=10). *Lebanese Ministry of Tourism.* . Retrieved 19 October 2009.

[148] Smith, Whitney (2009). "Lebanon, flag of" (http://www.britannica.com/EBchecked/topic/1355352/flag-of-Lebanon). *Encyclopædia Britannica.* . Retrieved 10 August 2009.

[149] Bell, Bethany (27 August 2008). "Threat to Lebanon's symbol of survival" (http://news.bbc.co.uk/2/hi/middle_east/7583757.stm). BBC. . Retrieved 10 August 2009.

[150] "Government" (http://www.lebanonembassy.nl/government.htm). *Embassy of Lebanon to the Netherlands.* . Retrieved 10 August 2009.

[151] Carter, Terry; Dunston Lara (15 July 2008). "Arts" (http://books.google.com/?id=_R-I_Gx5OgQC&pg=PA255). *Lonely Planet Syria & Lebanon.* Lonely Planet. Thomas Amelia (3 ed.). Lonely Planet. pp. 254–255. ISBN 978-1741046090. . Retrieved 19 September 2009.

[152] Sheehan, Sean; Latif Zawiah (30 August 2007). "Arts" (http://books.google.com/?id=cA-RDzlwVVAC&pg=PA105). *Lebanon.* Cultures of the World (2 ed.). Marshall Cavendish Children's Books. p. 105. ISBN 978-0761420811. . Retrieved 19 September 2009.

[153] McKenzie, Robert. *Comparing Media from Around the World*, Pearson/Allyn and Bacon, 2006, p. 372 ISBN 0205402429

[154] Kamalipour, Yahya; Rampal Kuldip (15 November 2001). "Between Globalization and Localization" (http://books.google.com/ ?id=yL3l0GwdNcsC&pg=PA265). *Media, sex, violence, and drugs in the global village.* Rowman & Littlefield Publishers, Inc.. p. 265. ISBN 978-0742500617. . Retrieved 19 September 2009.

[155] One source says "cult following", other says "folk hero"

[156] World Intellectual Property Organization (2003). "Copyright Industries in Lebanon" (http://books.google.com/?id=BzygcVYOpa8C& pg=PA148). *Performance of copyright industries in selected Arab countries: Egypt, Jordan, Lebanon, Morocco, Tunisia.* World Intellectual Property Organization. pp. 148–152. ISBN 978-9280513165. . Retrieved 19 September 2009.

[157] Karam, Michael (27 October 2005). *Wines of Lebanon* (http://nowlebanon.com/Sub.aspx?ID=173&MID=24&PID=23& FParentID=3&FFParentID=38). Saqi Books. p. 263. ISBN 978-0863565984. . Retrieved 18 September 2009.

[158] Christopher Reed Stone (2008). *Popular culture and nationalism in Lebanon: the Fairouz and Rahbani nation* (http://books.google.com/ books?id=-EdxrNmDE1UC&pg=PA50). Taylor & Francis. pp. 50–. ISBN 978-0-415-77273-0. . Retrieved 11 December 2011.

[159] Migliorino, p. 122

[160] "Lebanon profile – Overview" (http://news.bbc.co.uk/2/hi/middle_east/country_profiles/791071.stm). BBC News. 2011-08-24. . Retrieved 2011-11-04.

[161] Dale F. Eickelman; Jon W. Anderson (1 July 2003). *New media in the Muslim world: the emerging public sphere* (http://books.google. com/books?id=Moh2l5d85OYC&pg=PA63). Indiana University Press. pp. 63–65. ISBN 978-0-253-34252-2. . Retrieved 11 December 2011.

[162] Roy Armes (23 August 2010). *Arab filmmakers of the Middle East: a dictionary* (http://books.google.com/books?id=rHT8LjR_kC4C& pg=PA26). Indiana University Press. pp. 26–. ISBN 978-0-253-35518-8. . Retrieved 11 December 2011.

[163] Harabi, Najib (University of Applied Sciences, Northwestern Switzerland) Knowledge Intensive Industries: Four Case Studies of Creative Industries in Arab Countries (http://info.worldbank.org/etools/docs/library/251761/day3creative Industry WB_version 1.pdf), World Bank Project (May 2009) p. 16.

[164] Arabic and the Art of Printing: A Special Section (http://www.saudiaramcoworld.com/issue/198102/arabic.and.the.art.of.printing-a. special.section.htm). Saudi Aramco World. Retrieved on 2011-12-11.

[165] The First Arabic Script Printing Press in Lebanon: Arabic Type Designer & Typographer: Arabic Type: Pascal Zoghbi (http://29letters.
       wordpress.com/2009/01/05/the-first-arabic-press/). 29letters.wordpress.com (2009-01-05). Retrieved on 2011-12-11.

[166] *Lebanon A Country Study* by Federal Research Division, p. 42

[167] Andrew Hammond (2005). *Pop culture Arab world!: media, arts, and lifestyle* (http://books.google.com/books?id=m4eodEw7ZvsC&
       pg=PA94). ABC-CLIO. pp. 94–. ISBN 978-1-85109-449-3. . Retrieved 11 December 2011.

[168] Migliorino, p. 123

[169] Migliorino, p. 124

[170] Anker, Jean. *Libri: Volume 51*

[171] "Culture :: Books :: Francophone book fair showcases Lebanese and foreign authors" (http://www.dailystar.com.lb/Culture/Books/
       2011/Oct-28/152419-francophone-book-fair-showcases-lebanese-and-foreign-authors.ashx#axzz1c2rFfawB). The Daily Star. 2011-10-28. .
       Retrieved 2011-11-04.

[172] Zahera Harb (30 May 2010). *Channels of Resistance in Lebanon: Liberation Propaganda, Hezbollah and the Media* (http://books.google.
       com/books?id=p7FOdul0mbkC&pg=PA97). I.B.Tauris. pp. 97–. ISBN 978-1-84885-120-7. . Retrieved 11 December 2011.

References

Bibliography

- Migliorino, Nicola (2008). *(Re)constructing Armenia in Lebanon and Syria: ethno-cultural diversity and the state in the aftermath of a refugee crisis* (http://books.google.com/books?id=y_Sd32i-0owC&pg=PA166). Berghahn Books. ISBN 978-1-84545-352-7. Retrieved 11 December 2011.
- Morris, Benny (April 2008). *1948: A History of the First Arab-Israeli War* (http://books.google.com/?id=CC7381HrLqcC&dq=1948:+A+History+of+the+First+Arab-Israeli+War&printsec=frontcover&q=). Yale University Pres. ISBN 978-0300126969.

Further reading

- Arkadiusz, Plonka. *L'idée de langue libanaise d'après Sa'īd 'Aql*, Paris, Geuthner, 2004 (French) ISBN 2-7053-3739-3
- Firzli, Nicola Y. *Al-Baath wa-Lubnân* [Arabic only] ("The Baath and Lebanon"). Beirut: Dar-al-Tali'a Books, 1973
- Fisk, Robert. *Pity the Nation: The Abduction of Lebanon.* New York: Nation Books, 2002.
- Glass, Charles, "Tribes with Flags: A Dangerous Passage Through the Chaos of the Middle East", Atlantic Monthly Press (New York) and Picador (London), 1990 ISBN 0-436-18130-4
- Hitti Philip K. *History of Syria Including Lebanon and Palestine, Vol. 2* (2002) (ISBN 1-931956-61-8)
- Holst, Sanford. *Phoenicians: Lebanon's Epic Heritage.* Los Angeles: Cambridge and Boston Press, 2005.
- Norton, Augustus R. *Amal and the Shi'a: Struggle for the Soul of Lebanon.* Austin and London: University of Texas Press, 1987.
- Sobelman, Daniel. New Rules of the Game: Israel and Hizbollah After the Withdrawal From Lebanon, Jaffee Center for Strategic Studies, Tel-Aviv University, 2004.
- Riley-Smith, Jonathan. *The Oxford Illustrated History of the Crusades.* New York: Oxford University Press, 2001.
- Salibi, Kamal. *A House of Many Mansions: The History of Lebanon Reconsidered.* Berkeley: University of California Press, 1990.
- Schlicht, Alfred. The role of Foreign Powers in the History of Syria and Lebanon 1799–1861 in: Journal of Asian History 14 (1982)

External links

- Lebanon (http://www.informs.gov.lb/informs_en/pages/home.aspx) *official government portal*
- Lebanon (https://www.cia.gov/library/publications/the-world-factbook/geos/le.html) entry at *The World Factbook*
- Lebanon (http://ucblibraries.colorado.edu/govpubs/for/lebanon.htm) web resources provided by GovPubs at the University of Colorado–Boulder Libraries
- Lebanon (http://www.arabdecision.org/coun_sel_3_4.htm.htm) profiles of people and institutions provided by the *Arab Decision* project
- Lebanon (http://www.dmoz.org/Regional/Middle_East/Lebanon/) at the Open Directory Project
- Wikimedia Atlas of Lebanon
- Lebanon travel guide from Wikitravel

kbd:Ливан

Gebran_Tueni

Gebran Tueni	
Born	September 15, 1957Achrafieh, Beirut
Died	December 12, 2005Mkalles, Matn District
Cause of death	Assassination
Resting place	Saint Dimitrius church
Nationality	Lebanese
Occupation	Journalist, Politician
Political party	Qornet Shehwan Gathering
Religion	Greek Orthodox
Spouse	Mirna Murr (divorced) Siham Asseily
Children	Nayla Tueni(b. 1982) Michelle (b. 1987) Gabrielle (b. 2005) Nadia (b. 2005)
Parents	Ghassan Tueni Nadia Hamadeh
Relatives	Marwan Hamadeh (Uncle)
Website	
http://www.gebrantueni.com	

Gebran Ghassan Tueni (Arabic:) (September 15, 1957 – December 12, 2005) was a Lebanese politician and the former editor and publisher of the mass circulation An-Nahar daily newspaper in Beirut, Lebanon.

Tueni was a third generation journalist. An-Nahar was established by his grandfather, also named Gebran Tueni, in 1933. His father, Ghassan Tueni, ran the newspaper for decades.

Tueni had degrees in journalism, international relations and management from French universities. His mother was the famous Francophone, Lebanese Druze poet, Nadia Hamadeh Tueni. His uncle was the Druze, Telecommunications Minister Marwan Hamadeh.

Tueni came to international prominence in March 2000 when he wrote an editorial calling for the withdrawal of Syrian troops from Lebanon.[1] In March 2005, he contributed to the Cedar Revolution demonstrations during which he gave the famous "In the name of God We, Muslims and Christians, Pledge that united we shall remain to the end of time to better defend our Lebanon" speech. In May 2005 he was elected a member of Parliament of Lebanon for the Greek Orthodox seat in Beirut on an anti-Syrian slate led by Saad al-Hariri, son of assassinated former Prime Minister Rafik al-Hariri. He was a member of the political coalition Qornet Shehwan Gathering headed by Catholic Maronite bishop Youssef Bechara.

Assassination

The member of the Parliament Tueni was assassinated by a car bomb on December 12, 2005 in Mkalles, an industrial suburb of Beirut, while on his way to work. He has been buried at Saint Dimitrius church graves after the funeral that took place at Saint George church Beirut. Initial reports indicated that a hitherto unknown group, "Strugglers for the Unity and Freedom of al-Sham" (where al-Sham refers to ancient Greater Syria) claimed responsibility. The statement taking responsibility was faxed to Reuters and included a warning that the same fate awaited other opponents of "Arabism" in Lebanon, claiming that the assassination has succeeded in "shutting up" a traitor, and "turning An-nahar" (Arabic for Day) into Dark Night.

Tueni's assassination coincided with the release of the second progress report of a United Nations inquiry into Syria's involvement in the assassination of Rafik Hariri. In response, Lebanese Prime Minister Fouad Siniora announced that he would ask the United Nations Security Council to investigate Syrian complicity in the deaths of Tueni and other prominent anti-Syrian figures.

Before his death, Tueni was campaigning for an international probe into recently discovered mass graves in Anjar next to the main Syrian intelligence headquarters. Forensic analysis later showed the graves were part of an 18th century cemetery. In his last editorial Tueni accused Syria of committing "crimes against humanity" and blamed them for the mass graves and other atrocities committed in Lebanon during their presence. His articles and editorials in An-Nahar often raised the ire of the Syrians.

Tens of thousands of mourners filled the streets of Beirut for Tueni's funeral on 14 December 2005. Many mourners blamed Syria for his death due to his antisyrian policy and they chanted anti-Syrian slogans. Members of the Lebanese parliament also observed a moment of silence during a special parliamentary session. Continuing the play on words with "An-nahar" (*The Day*), family members stated that night would not fall on the newspaper.

Gibran died at the age of 48 leaving a widow and 2 newly born daughters along with 2 other daughters from a previous marriage.

References

[1] Gibran Tueni: Open Letter to Bashar Assad (http://www.meib.org/articles/0004_doc1.htm)

External links

- http://www.gebrantueni.com

Lebanese_Forces

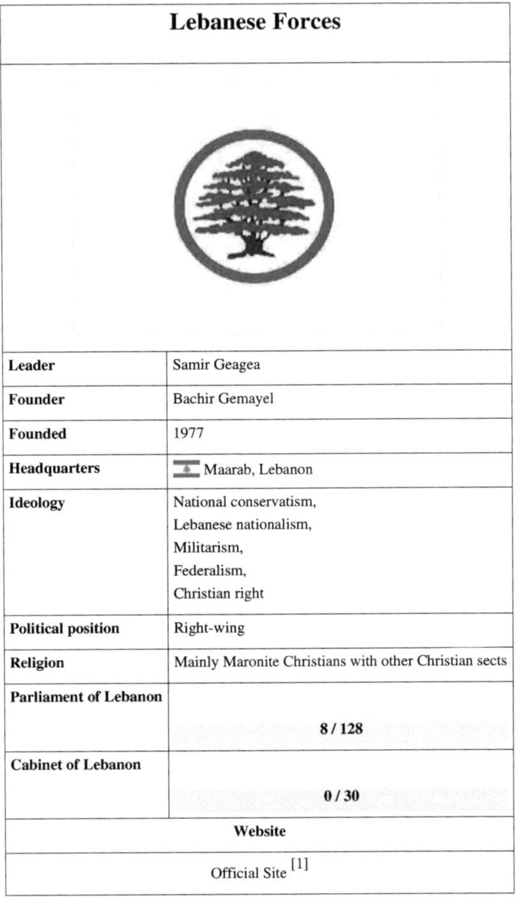

Lebanese Forces	
Leader	Samir Geagea
Founder	Bachir Gemayel
Founded	1977
Headquarters	Maarab, Lebanon
Ideology	National conservatism, Lebanese nationalism, Militarism, Federalism, Christian right
Political position	Right-wing
Religion	Mainly Maronite Christians with other Christian sects
Parliament of Lebanon	8 / 128
Cabinet of Lebanon	0 / 30
Website	
Official Site [1]	

The **Lebanese Forces** (LF) (Arabic: القوات اللبنانية *al-quwāt al-lubnāniyah*, Syriac: ܚܝܠܘܬܐ ܠܒܢܢܝܐ *ḥailaoṭe lebnonoye*) is a Lebanese political party. Founded as a militia by Bachir Gemayel during the Lebanese Civil War, the movement fought as the main militia within the Christian-dominated Lebanese Front.[2] The militia mainly fought the militants of the Palestinian Liberation Organization and the Syrian troops occupying Lebanon.

After the civil war ended, the movement reinvented itself as a political party under the leadership of Samir Geagea. In 1994, while Lebanon was under Syrian occupation the party was banned, Geagea imprisoned, and the activities of its militants repressed by the Lebanese services in Lebanon. The Lebanese Forces returned as a political force after the Cedar Revolution in early 2005, which resulted in a withdrawal of Syrian troops from Lebanon. Soon after, Geagea was subsequently released from prison and continues to lead the party today.

Formation

The Lebanese Front was informally organized in January 1976 under the leadership of Bashir's father, Pierre Gemayel and Camille Chamoun. It began as a simple coordination or joint command between the predominantely Christian Kataeb Party/Kataeb Regulatory Forces (KRF), Tyous Team of Commandos (TTC), Ahrar/Tigers Militia, Al-Tanzim, Marada Brigade and Lebanese Renewal Party/Guardians of the Cedars (GoC) parties and their respective military wings. The main reason behind the formation of the Lebanese Front was to strengthen the Christian side against the challenge presented by the Lebanese National Movement (LNM), an umbrella alliance of leftist Muslim parties/militias backed by the Palestine Liberation Organization (PLO) and Rejectionist Front Palestinian guerrilla factions.

The Golden Years (1976-1982)

Christian East Beirut was ringed by heavily fortified Palestinian camps from which kidnappings and sniping against Lebanese civilians became a daily routine. Christian East Beirut became besieged by the PLO camps, with severe shortages of food and fuel. This unbearable situation was remedied by the Kataeb Regulatory Forces (most notably the BG Squad that was led by Bachir) and their allied Christian militias as they besieged the Palestinian camps embedded in Christian East Beirut one at a time and brought them down. The first was on 18 January 1976 when the heavily fortified Karantina camp, located near the strategic Beirut Harbor, was invaded: About 1,000 PLO fighters and civilians were killed.[3] The Palestinian PLO and al-Saiqa forces retaliated by attacking the isolated defenseless Christian town of Damour about 20 miles south of Beirut on the coast, during the Damour massacre in which 1,000 Christian civilians were killed and 5,000 were sent fleeing north by boat, since all roads were blocked off.[4] The Maronites retaliated with the invasion of the largest and strongest Palestinian refugee camp, Tel al-Zaatar that same year.[5] Bachir, with his KRF militia units, also fought against the PLO and LNM militias at the Battle of the Hotels in central Beirut. The most important battle won by the Phalange for the control of the hotel district was the fighting over the possession of the Holiday Inn, due to its important strategic location. Before that battle, the Holiday Inn had been occupied by the PLO.[6]

Relations within the Lebanese Front between militias started to decline when the Marada party, led by Tony Frangieh and the Phalange, whose military wing was now led by Bachir Gemayel, began to clash. On June 13, 1978, Bachir sent a group of Phalangists to Ehden. The aim of this operation was to kidnap the members of the Marada Militia responsible for the killing of many Phalangists in the North. The operation was postponed till Tuesday to make sure that Tony Frangieh would have finished his weekend vacation in Ehden; however, his car did not work and he stayed in Ehden. When the Phalangists arrived to Ehden, bullets were flying above their heads, so they retaliated and they ended up disobeying their orders, killing Frangieh, his family and militiamen in the Ehden Massacre.[7] This, as well as other problems within the Lebanese Front, caused the group to collapse.

The Lebanese Forces was soon after established with an agreement that the direct military commander would be a Kataeb member and the vice-commander an Ahrar member.

Bachir led his troops in the infamous "Hundred Days War" in Lebanon in 1978, in which the Lebanese Forces successfully resisted the Syrian shelling and attacking of Eastern Beirut for about three months before an Arab-brokered agreement forced the Syrians to end the siege. Syrians took high buildings such as Burj Rizk Achrafieh and Burj El Murr using snipers and heavy weapons against civilians. The soldiers stayed for 90 days. Another major clash took place near the Sodeco area in Achrafieh where the Lebanese Forces fought ferociously and led the Syrian army out of the Rizk Building.[8] At this time, Israel was the primary backer of the Lebanese Front's militia.

In July 1980, following months of intra-Christian clashes between the Tigers, the militia of Dany, and the Phalangists, who by now were under the complete leadership of Bachir Gemayel, the Phalangists launched an operation in an attempt to stop the clashes within the Christian areas, and to unite all the Christian militias under

Gemayel's command. This operation resulted in a massacre of tens of Tigers' members at the Marine beach resort in Safra, 25 km north of Beirut. Camille Chamoun's silence was interpreted as acceptance of Gemayel's controls, because he felt that the Tigers led by his son were getting out of his control.[9]

In 1981 at Zahlé in the Beqaa, the largest Christian town in the East, confronted one of the biggest battles – both military and political – between the Lebanese Forces and the Syrian occupying forces. The Lebanese Forces was able to confront them even though there was a big mismatch in military capabilities and was able to reverse the result of the battle of 1981. This victory was due to the bravery of the inhabitants and 92 Lebanese Forces soldiers (L.F Special Forces: The Maghaweer) sent from Beirut. The Syrian occupying forces used all kind of weapons (heavy artillery, tanks, war planes...) against a peaceful town, and they cut all kind of backup that may come from the Mountain. Regardless of the very bad weather and heavy bombing, convoys were sent in the snow to Zahle. Two Lebanese Forces soldiers died on a hill due to bad weather, they were found later holding each other... till they died. The battle of Zahle gave the Lebanese Cause a new perspective in the International Communities, and the victory was both military and diplomatic. It made the Leadership of President Bashir Gemayel much stronger because of his leadership and important role in this battle. The battle started in April the 2nd 1981, and finished with a cease fire and Lebanese Police were sent to Zahle. The 92 Lebanese Forces heroes returned to Beirut on the 1st of July 1981.[10]

Israeli Invasion

In 1982, Bachir met with Hani Al-Hassan (representative of the PLO) and told him that Israel will enter and wipe them out. Bachir told him to leave Lebanon peacefully before it's too late. Hani left and no reply was given to Bachir.[11]

Israel invaded Lebanon, arguing that a military intervention was necessary to root out PLO guerrillas from the southern part of the country. Israeli forces eventually moved towards Beirut and laid siege to the city, aiming to reshape the Lebanese political landscape and force the PLO out of Lebanon. By 1982, Israel had been the main supplier to the Lebanese Forces, giving them assistance in weapons, clothing, and training.

After the PLO had been expelled from the country to Tunisia, in a negotiated agreement, Bachir Gemayel became the youngest man to ever be elected as president of Lebanon. He was elected by the parliament in August; most Muslim members of parliament boycotted the vote.

On September 3, 1982, During the meeting, Begin demanded that Bachir sign a peace treaty with Israel as soon as he took office in return of Israel's earlier support of Lebanese Forces and he also told Bachir that the IDF will stay in South Lebanon if the Peace Treaty was not directly signed. Bachir was furious at Begin and told him that the Lebanese Forces did not fight for seven years and that they did not sacrifice thousands of soldiers to free Lebanon from the Syrian Army and the PLO so that Israel can take their place. The meeting ended in rage and both sides were not happy with each other.[12]

Begin was reportedly angry at Bachir for his public denial of Israel's support. Bachir refused the immediate peace arguing that time is needed to reach consensus with Lebanese Muslims and the Arab nations. Bachir was quoted telling David Kimche, the director general of the Israeli Foreign Ministry, few days earlier, "Please tell your people to be patient. I am committed to make peace with Israel, and I shall do it. But I need time - nine months, maximum one year. I need to mend my fences with the Arab countries, especially with Saudi Arabia, so that Lebanon can once again play its central role in the economy of the Middle East."[13] [14]

In an attempt to fix the relations between Bachir and Begin, Ariel Sharon met secretly with Bachir in Bikfaya. In this meeting, they both agreed that, after 48 hours, the IDF will cooperate with the Lebanese Army to force the Syrian Army out of Lebanon. After that is done, the IDF would peacefully leave the Lebanese territory. Concerning the Peace Negotiation, Sharon agreed to give Bachir time to fix the internal conflicts before signing the negotiation. The next day, Begin's office issued a statement saying that the issues agreed upon between Bachir and Sharon were accepted.[15]

Nine days before he was to take office, on September 14, 1982, he was killed along with 25 others in a bomb explosion in the Kataeb headquarters in Achrafieh. The attack was carried out by Habib Shartouni, a member of the Syrian Social Nationalist Party (SSNP), believed by many to have acted on instructions of the Syrian government of President Hafez al-Assad.[16] The next day, Israel moved to occupy the city, allowing Phalangist members under Elie Hobeika's command to enter the centrally located Palestinian refugee camps of Sabra and Shatila; a massacre followed, in which Phalangists killed 400 Palestinian refugees, causing great international uproar. Many cite the massacre as revenge for the killing of Bachir Gemayel and the countless massacres committed by the PLO against the Christian civilian population since 1975.

The Amine Gemayel years (1982–1988)

Battles

Mountain War

After the Israeli invasion, the IDF troops settled in the Chouf and Aley districts in Mount Lebanon. The Lebanese Forces returned to the Christian villages which had been occupied by the PSP for seven years, and many Christian civilians from the districts returned after having fled earlier in the war. However, soon after, clashes broke out between the Lebanese Forces and the Druze militias who had now taken over the districts and had earlier kicked out the Christian inhabitants. The main Druze militiamen came from the Progressive Socialist Party, led by Walid Jumblatt, in alliance with the Syrian Army and Palestinian militants who had not departed Lebanon in 1982. For months, the two fought what would later be known as the "Mountain War." At the peak of the battle, Israeli troops infamously abandoned the area, handing over the best tactical positions to the Druze militias and their allies as punishment for the Christians' refusal of the May 17 peace agreement with Israel, and leaving the Christian forces sitting ducks ready to be slaughtered. Even though the Christian inhabitants of these regions were almost entirely with Jumblatt's PSP, and historically very loyal to Kamal Jumblatt, more than two thousand Christian civilians were massacred in the ensuing invasion, most of whom were killed after surrendering, where Druze would conduct the mssacres with almost medieval style weapons, and their Palestinian and Syrian allies would do most of the fighting. The total destruction of tens of villages, towns, churches and monasteries ensured the complete extermination of the millennium old Christian mountain population.

Ironically, the Palestinian militants, the Christian's main enemies in the war, helped save countless civilian lives by going from town to town and warning the hapless civilians that the Druze militias were advancing and bent on killing them all, giving them enough time to flee the mountain.

The massacre is estimated to be the largest of the Lebanese war, and had reached almost genocidal proportions.

At the same time, a small number of ill equipped Lebanese Forces troops also fought battles against the Palestinian and Druze militias and the Syrian troop east of the southern city of Sidon. The outcome was also a Progressive Socialist Party victory and a contiguous Druze Chouf district with access to Lebanese sea ports.

Jumblatt's militia then overstepped itself by attacking further into Souk El Gharb, a village held by the Lebanese Army's multi-confessional 8th Mechanised Infantry commanded by Army Chief Michel Aoun. The attackers were fiercely pushed back.

Internal power struggles

After the death of Bachir, his brother Amine Gemayel replaced him as President, and his cousin, Fadi Frem as commander of the Lebanese Forces. The two had a frosty relationship, and in 1984, pressure from Amine led to Frem's replacement by Fouad Abou Nader.

On March 12, 1985, Samir Geagea, Elie Hobeika and Karim Pakradouni rebelled against Abou Nader's command, ostensibly to take the Lebanese Forces back to its original path. The relationship between Geagea and Hobeika soon broke down, however, and Hobeika began secret negotiations with the Syrians. On December 28, 1985, he signed the Tripartite Accord, against the wishes of Geagea and most of the other leading Christian figures. Claiming that the Tripartite Accord gave Syria unlimited power in Lebanon, Geagea mobilized factions inside the Lebanese Forces and on January 15, 1986, attacked Hobeika's headquarters in Karantina. Hobeika surrendered and fled, first to Paris and subsequently to Damascus, Syria. He then moved to Zahlé with tens of his fighters where he prepared for an attack against East Beirut. On September 27, 1986, Hobeika's forces tried to take over the Achrafieh neighborhood of Beirut but the Lebanese Forces of Geagea's command held them back.

This failed attempt by Hobeika was the last episode of internal struggles in East Beirut during Amine Gemayel's mandate. As a result, the Lebanese Forces led by Geagea were the only major force on ground. During two years of frail peace, Geagea launched a drive to re-equip and reorganize the Lebanese Forces. He also instituted a social welfare program in areas controlled by Geagea's party. The Lebanese Forces also cut its relations with Israel and emphasized relations with the Arab states, mainly Iraq but also Saudi Arabia, Jordan, and Egypt.

The Elimination War (1988–1990)

Two rival governments contended for recognition following Amine Gemayel's departure from the Presidency in September 1988, one a mainly Christian government and the other a government of Muslims and Lebanese Leftists. The Lebanese Forces initially supported the military Christian government led by Gen. Michel Aoun, the commander of the Lebanese Army. However, clashes erupted between the Lebanese Forces and the Lebanese Army under the control of Michel Aoun on February 14, 1989. These clashes were stopped, and after a meeting in Bkerké, the Lebanese Forces handed the national ports which it controlled to Aoun's government under pressure from the Lebanese National army.

Captain Salim Meayki was killed during attempted peace negotiations with the Lebanese Army.

Geagea initially supported Aoun's "Liberation War" against the Syrian army, but then agreed to the Taif Agreement, which was signed by the Lebanese deputies on 24 October 1989 in Saudi Arabia and demanded an immediate ceasefire. Aoun's main objection to the Taif Agreement was its vagueness as to Syrian withdrawal from the country. He rejected it vowing that he "would not sign over the country." Fierce fighting in East Beirut broke out between the two, called the "Elimination War" on January 31, 1990.

The Second Republic (1990–2005)

After Aoun surrendered on 13 October 1990 to the rival Syrian-backed President Hrawi, Geagea was offered ministerial posts in the new government. He refused several times, because he was opposed to Syrian interference in Lebanese affairs, and his relationship with the new government deteriorated. On March 23, 1994, the Lebanese government ordered the dissolution of the LF.[17] On April 21, 1994, Geagea was arrested on charges of setting a bomb in the church in Zouk, of instigating acts of violence, and of committing assassinations during the Lebanese Civil War. Although he was acquitted of the first charge, Geagea was subsequently arrested and sentenced to life imprisonment on several different counts, including the assassination of former Prime Minister Rashid Karami in

1987. He was incarcerated in solitary confinement, with his access to the outside world severely restricted. Amnesty International criticized the conduct of the trials and demanded Geagea's release, and Geagea's supporters argued that the Syrian-controlled Lebanese government had used the alleged crimes as a pretext for jailing Geagea and banning an anti-Syrian party. Many members of the Lebanese Forces were arrested and brutally tortured in the period of 1993-1994. At least one died in Syrian custody and many others were severely injured.[18]

After the Cedar Revolution

The LF was an active participant in the Cedar Revolution of 2005, when popular protests and international pressure following the assassination of former Prime Minister Rafiq al-Hariri combined to force Syria out of Lebanon. In the subsequent parliamentary election held in May and June, the Lebanese Forces formed part of the Rafik Hariri Martyr List, which also included the Future Movement, Popular Socialist Party, the reformed Phalange party, and other anti-Syrian political groups, as well as a brief tactical alliance with Amal and Hezbollah. The tactical alliance with Hizbollah and Amal would soon end ; these majority parties and movements would subsequently form the anti-Syrian March 14 Alliance, which stood opposed to the March 8 Coalition backed by Hizbullah, Amal and the Free Patriotic Movement led by General Michel Aoun who had returned to Lebanon. The Lebanese Forces were able to win 6 out of the 8 MPs that were nominated throughout the various regions of the country. Nevertheless, the elections proved to be very significant because for the first time, supporters of the party were freely able to participate in the election process.

Following the party's new political gains, Samir Geagea was freed on 18 July 2005, after parliament decided to amend all the charges he formerly faced. Since Geagea's release from prison, the Lebanese Forces have been rebuilding much of their former image. Some of these works include reorganizing its members and their families, reopening political facilities, and reestablishing their main presence among the Christians of Lebanon. In addition to rebuilding their image, the Lebanese Forces have also been attempting to reclaim former privately-funded facilities, which were seized by the Syrian backed government. Currently, the Lebanese Forces have also been striving to reclaim their rights to the Lebanese Broadcasting Corporation, which was initiated by the party in the mid 1980s. After filing suit against LBC Group General Manager Pierre Daher in 2007, the Lebanese Forces won the case and were granted control of the corporation in late 2010.

Since the emancipation of the party's main leader, Samir Geagea, the party has gained new popularity among the Christian population throughout all of Lebanon. In addition, the Lebanese Forces have also been able to attain a great deal of popularity amongst the younger generation, as evidenced by the annual student elections in Lebanese colleges. The Lebanese Forces, along with their other March 14 allies, made additional gains in the elections geared towards the professional bodies of engineers, doctors, lawyers, and even teachers.

Present Political Representation

The Lebanese Forces currently hold 8 out of the 128 seats of the Lebanese Parliament, and were represented in the Siniora government, formed in July 2005, by the minister of Tourism Joseph Sarkis, and then in the second Siniora government, formed in July 2008, by the minister of Justice Ibrahim Najjar and the minister of Environment Antoine Karam. They are the leading Christian party within the March 14 Bloc, an anti-Syrian movement.

Today, the Lebanese Forces and its main political representatives strive to re-establish the many Christian rights, which were significantly lessened during Syria's occupation of Lebanon, specifically from 1990-2005. Some of the Lebanese Force's other main objectives include formulating a just electoral law, which would enable the Christian population to be represented fairly in local and parliamentary elections. The party also strives to give the large Lebanese diaspora the ability to participate in Lebanese elections. The party has also stressed the idea of reaffirming the powers formerly endowed to the Lebanese president before being lessened in the Taef Agreement.

Leaders

- Bachir Gemayel (1978–1982)
- Fadi Frem (1982–1984)
- Fouad Abou Nader (1984–1985)
- Elie Hobeika (1985–1986)
- Samir Geagea (1986–present)

Current Deputies

- Georges Adwan - Elected in 2005, re-elected in 2009.
- Elie Kayrouz - Elected in 2005, re-elected in 2009.
- Antoine Zahra - Elected in 2005, re-elected in 2009.
- Farid Habib - Elected in 2005, re-elected in 2009.
- Sethrida Geagea - Elected in 2005, re-elected in 2009.
- Toni Abi Khater - Elected in 2009.
- Joseph Maalouf - Elected in 2009.
- Chant Jinjenian - Elected in 2009.

See also

- Bachir Gemayel
- Samir Geagea
- Lebanese Civil War
- Lebanese Forces – Executive Command
- Cedar Revolution
- Political parties in Lebanon
- Kataeb Regulatory Forces
- Tyous Team of Commandos

References

[1] http://www.lebanese-forces.com

[2] "1978: Israeli troops leave southern Lebanon" (http://news.bbc.co.uk/onthisday/hi/dates/stories/june/13/newsid_2512000/2512241. stm). *BBC*. June 13, 1978. . Retrieved 2008-01-17.

[3] Harris (p. 162) notes "the massacre of 1,500 Palestinians, Shi'is, and others in Karantina and Maslakh, and the revenge killings of hundreds of Christians in Damur"

[4] http://www.lebaneseforces.com/blastfromthepast002.asp

[5] http://www.liberty05.com/civilwar/civil1.html

[6] "LEBANON: Beirut's Agony Under the Guns of March" (http://www.time.com/time/magazine/article/0,9171,913978,00.html). *Time*. April 5, 1976. .

[7] http://www.ouwet.com/n10452/political/ehden-1978-what-happened-there/

[8] http://www.youtube.com/watch?v=t28UYcXZRCs

[9] http://en.academic.ru/dic.nsf/enwiki/1200819

[10] http://www.lebaneseforces.com/blastfromthepast009.asp

[11] قصة المواورنة في الحرب - جوزيف أبو خليل

[12] تاريخ في رجل --- من قتل بشير - إنقلاب بشيري أم إنقلاب إسرائيلي

[13] *President Reagan and the World* by Eric J. Schmertz, Natalie Datlof, Alexej Ugrinsky, Hofstra University

[14] Special to the New York Times (1982-09-04). "Begin Said to Meet in Secret With Beirut's President-Elect". The New York Times. "Begin Said to Meet in Secret With Beirut's President-Elect"

[15] أسرار الحرب في لبنان

[16] "Phalangists Identify Bomber Of Gemayel As Lebanese Leftist" (http://query.nytimes.com/gst/fullpage. html?res=9E0DE0DB1F38F930A35753C1A964948260). *The New York Times*. October 3, 1982. . Retrieved May 7, 2010.

[17] Lebanon Detains Christian in Church Blast (http://query.nytimes.com/gst/fullpage.
      html?res=9B06E6D8133CF937A15750C0A962958260&scp=2&sq=geagea&st=nyt). *New York Times*, March 24, 1994. Retrieved on
      2008-02-13.

[18] UN Commission on Human Rights - Torture - Special Rapporteur's Report (http://www1.umn.edu/humanrts/commission/thematic51/
      34.htm). *United Nations Economic and Social Council*, January 12, 1995. Retrieved on 2008-03-06.

External links

- Lebanese Forces Official Website (http://www.lebanese-forces.org)
- Lebanese Forces Forum (http://www.lebanese-forces.org/forum/index.php)
- Lebanese Forces Martyrs Website (http://www.lfmartyrs.org)
- Lebanese Forces Television (http://lftv.tv/)
- Lebanese Forces Blog (http://www.ouwet.com)

Article Sources and Contributors

Nayla_Tueni *Source*: http://en.wikipedia.org/w/index.php?title=Nayla_Tueni *Contributors*: Aboutmovies, Arab Hafez, Gidonb, Good Olfactory, LilHelpa, Majp, Miss-simworld, Quoniraat, Rolatee, SpaceFlight89, Supreme Deliciousness, Tim1357, Traveller360, WikHead, Xezbeth, 22 anonymous edits

Ghassan_Tueni *Source*: http://en.wikipedia.org/w/index.php?title=Ghassan_Tueni *Contributors*: Al7akim79, Caiaffa, CharlieDelta, Elie plus, Fram, Good Olfactory, Hiram111, JYi, Jeodesic, KC Panchal, Klaksonn, Louaymk, Miss-simworld, Quoniraat, R'n'B, Tellyaddict, The Thing That Should Not Be, Topbanana, Waacstats, Werldwayd, With14ever, Zalali, 17 anonymous edits

Politician *Source*: http://en.wikipedia.org/w/index.php?title=Politician *Contributors*: 123bubble, 71Demon, AbigailAbernathy, Acather96, AdamRetchless, Ae00111, Aldo, Alexius08, Alligators1974, AlphaEta, Altenmann, AmericanCentury21, Amitmgr, AndrewM1, Andy120290, Andycjp, Antonio Lopez, Aqwis, Arakunem, Artblogs, Auntof6, Auroranorth, Avoided, BesselDekker, Betterusername, Bfigura's puppy, Big Bird, Big Black shiny revenge, Binadot, Binksternet, Blue-Haired Lawyer, BlueMoonlet, Bojanglesk8, Briaboru, Brian0918, Brookie, CIreland, Calvin 1998, Capricorn42, Causa sui, Chezifresh, Chuunen Baka, CloutierFan02, CordeliaNaismith, Courcelles, Curufinwe, Cybersquire, D, DO'Neil, Danceswithzerglings, Danger, Daofktr, Ddye, DerHexer, Dfrg.msc, Dilane, Discospinster, Diyako, Docu, Dragonrider1015, Edcolins, Edison, Edrainkona, El diablo92, Electionworld, Ellywa, Elustran, Epbr123, EronMain, Eurovision222, Everyking, Excirial, Ezadarque, Falcon8765, Fatal-, Favonian, Femto, Fieldday-sunday, Francis Tyers, Frecklefoot, FreplySpang, Friginator, GH3roxx, Galoubet, Ggfre, Gianfranco, Giraffedata, Gnowor, Gråbergs Gråa Sång, Gscshoyru, Gurch, Haemo, HailFire, Hairy beast, HamburgerRadio, Happysailor, Hdt83, Helix84, Hirolovesswords, Hovden, Icyhandofcrap, IronGargoyle, Itai, J.delanoy, JForget, Ja 62, JackHearne, James dempsey, JamesBWatson, Jauc1back, Jaywheat, Jcuk, Jebba, Jerzy, Jiang, Jimothytrotter, Josh Parris, Jsmit12, Juan M. Gonzalez, Kaclock, Karenjc, Kate, Kbh3rd, Kendrick7, Kestenbaum, Kevlarsen, King of Hearts, Kingpin13, Kingturtle, Koert Vrijhof, Kokoriko, Kubigula, LOL, Larklight, LeaderOfTheBushRebellion, LeoO3, Lesterxyz, Leszek Jańczuk, Leuko, Levineps, Lightmouse, Lincolnite, Lkinkade, Lotje, Lou Sander, Luna Santin, Lupin, MER-C, Mairi, Mamainiero, Marek69, Mbch331, Mboverload, Meelar, Michaelshirley, Mihirbhojani, Mike Selinker, Millerlax19, Minesweeper, Mister bigglesworth, Mnbvcxzqa, Mobile Snail, Monty845, Moondyne, Morwen, Mrh30, Muchness, N419BH, N5iln, Nakos2208, NawlinWiki, Neutrality, Nibuod, NickelShoe, NorthernThunder, Nsaa, Numbo3, NurseryRhyme, Ojigiri, Oksana160773, Old Moonraker, Olpus, Oromanos, Orthoepy, Orzetto, P.AndersonWiki, PGSable, Palthian, PatGallacher, Patrolmanno9, Paul Klenk, Paul-L, Paul6077, Pce3@ij.net, Pgarwood, Pharaoh of the Wizards, PhilKnight, Philip Trueman, Phoenician Patriot, Phoyes, Phyrros, Piano non troppo, Pinethicket, Pizza Puzzle, Plastikspork, Politicalaffairs, Politicalheroes, Polly, Polycarp, Prashanthns, Prince Of Fire, Pronogos, Proofreader77, ProtomegaX, Prunesqualer, Puchiko, Qweqwewe, Qwyrxian, Qxz, RL, RadioFan, RainbowOfLight, Random contributor, RayAYang, Rayback, Reach Out to the Truth, Reaper Eternal, Retaggio, RexNL, Rich Farmbrough, Rintrah, Robert Merkel, Rojema, Roux, RoyBoy, SC, SD6-Agent, Safemariner, Sceptre, Sekhinov Alexey Victorovich, Seba, SebastianHawes, Secretlondon, Seeleschneider, Shoreranger, SimonP, Sionus, Skittleys, Slarson, Southern Illinois SKYWARN, Spencer1212, Steven J. Anderson, Striver, TVDATAOX, TakuyaMurata, Tamilnesan, Tasudrty, Taylortime17, Teles, Tempodivalse, TexasAndroid, Thaurisil, The Behnam, The Red Hat of Pat Ferrick, The Thing That Should Not Be, The sunder king, TheJazzDalek, Thingg, This.machinery, TicketMan, Tide rolls, TigerShark, Timberframe, Timeshift9, TitanOne, Tom harrison, TomCerul, ToniBirdie, Triwbe, Twooars, USchick, Ukbeast62696, Unclenuclear, Uttrillom17, VMS Mosaic, Valentinian, Vardion, Verybigfish86, Vsmith, Vzbs34, Wafulz, Weide, Wendell, Weregerbil, Whoopel, Wikidudeman, Wikijens, Wiktator, Willking1979, Windjade, WingateChristopher, Woohookitty, Xn4, Yekrats, Yintan, Zanaq, Zondor, 377 anonymous edits

Lebanese_University *Source*: http://en.wikipedia.org/w/index.php?title=Lebanese_University *Contributors*: Aboutmovies, Anonymous editor, Banzoo, Boulbass, CambridgeBayWeather, Canderson7, Captainm, Cedrus-Libani, Chankouteh, D6, Darwinek, Dsp13, EmanWilm, GreatWhiteNortherner, Grm wnr, Jaberwocky6669, Jaraalbe, Jengod, Kamyam, LebanonChild, MLeb, Masonpatriot, Ocaasi, Occitanstag, Omar 180, Orderinchaos 2, PRRfan, Rm uk, SimonP, Tyomitch, Whpq, Woohookitty, 21 anonymous edits

Journalist *Source*: http://en.wikipedia.org/w/index.php?title=Journalist *Contributors*: A Person 300, Addshore, Aeternus, Ageekgal, Ahaislip, Ahoerstemeier, Alai, Alfio, Alfonso Márquez, Anaxial, AndersL, Andyjsmith, Anonyme..., ArglebargleIV, Astuishin, Avaiki, Avoided, AzureCitizen, Backslash Forwardslash, Badanedwa, Banano03, BarryTheUnicorn, Basawala, Bebop, Benc, Birdmessenger, Bjh21, Bobblewik, Bobo192, Bodnotbod, Bougandoura, Branko, Brian0918, CDN99, Calicocat, Calton, CambridgeBayWeather, Camptown, Can't sleep, clown will eat me, Carpenoctem, Cgingold, Chaeyounggil, Champaignidea, Chatmanrace, CheesusChrist, ChildofMidnight, Chodorkovskiy, Christian List, Chuckino, Ckatz, Clindhartsen, ColumbiaMWD, Comelloyellow, CommanderBond, Conversion script, Craig Hicks, Cst17, CurvBall, Cyberprog, Cymi, Cyrius, Dan0707, Darkwind, DavidWBrooks, Davodd, Deep1979, Deficite, Dilipawasthi, Disavian, Dlyons493, Docu, Doldrums, Dougofborg, Dpalma01, Dumbass121, Dungodung, Dúnadan, E2eamon, Earthsky, Eclecticology, Edwin s, Elmer Clark, Elonka, EpicDream86, Erri4a, Escape Orbit, Espen, Everyking, Ex-User17, Favonian, Fazed554, Filippowiki, Flapdragon, Fluterst, FlyHigh, Francs2000, FreplySpang, Frugalnortherner, Fsotrain09, Fuzheado, Gdo01, GeorgeLouis, GeorgeOrr, Gimmetrow, Gladmax, Gnusmas, Gogo Dodo, GossamerBliss, Grafen, Graham87, GreatWhiteNortherner, Gwernol, Hajenso, Helixweb, Hetajatin, Hmains, IamBlack23101, ImNotBlue, Indon, Iranway, Isomorphic, JForget, JRR Trollkien, JTBurman, Jaberwocky6669, Jacklail, Jacobolus, Jarel75, Jay-W, JayJasper, Jayjg, Jeepday, Jeff3000, Jeremyburton, Jeroen Mirck, Jesslynn lee86, Jkaplan, Jliberty, Jmwalsh, Jmwoltjen, Johnbrownsbody, Jonemerson, JorgeGG, Joshua Issac, Joshuajohnlee, Joy, Joyous!, Jrtayloriv, KF, Kayau, Kbdank71, Kchishol1970, Kedar63, Kesla, Kingstonjr, Kingturtle, Kjbjournalism, KnowledgeOfSelf, Kurtbw, Kziabari, LaNicoya, Lavdo, Leaf of Silver, Lee Daniel Crocker, Lesterxyz, Lightmouse, Like tears in rain, LiveWire, Lokomotivewest, Lugnad, Luis Napoles, Lupo, MER-C, Maccess, Maelnuneb, Malcolm Farmer, Marj Tiefert, Master of Pies, Maurice Carbonaro, Maurreen, Maximus Rex, MayaSimFan, Mcbridelr, Meco, Mgordon42, Microchip08, Minesweeper, Mintguy, Mirwin, Mkamransaqi, Mkweise, Moverton, Mr. Billion, Mrutter, Muellerb18, Mwanner, N5iln, NaBURu38, Nakon, Neelix, Neutrality, Nihil novi, Nlu, Nminow, Nostalgicmat, Ocaasi, Oioioioioioioi, Ombudsman, Omnipaedista, Ortolan88, Otto4711, Outcast44, Oxymoron83, PBP, Palmiped, Paul A, Pha telegrapher, Philip Cross, Pilotguy, Pittimannl, Poitypoity, Pournami, Ppntori, Private Butcher, Puchiko, Pudmaker, Quercusrobur, R'n'B, RCEberwein, RabbitTrap, Raviamb, Rawdon, Rcr18, RedWolf, Reenem, Reevesg, Rexroad2, RickK, Rizla, Rkarnad, Rt Jones, Rtkat3, Russ0035, RustySpear, SMcCandlish, Samsara, Samulat, Savidan, Schekinov Alexey Victorovich, Scope creep, Secretlondon, Senator2029, ShakespeareFan00, Shorthandworld, Silversam, Simon123321, SimonP, Sjc, Skookum1, Slakr, Snackycakes, Sperril, Spitfire, SquidSK, StaticGull, Stephen pomes, Steven J. Anderson, Stevertigo, Stitchill, Subzer019, Susan Mason, SusanLesch, Swarm, TakuyaMurata, Tenmei, Texture, ThatPeskyCommoner, Theinsomniac4life, Tide rolls, Tobogganoggin, Tohd8BohaithuGh1, Tom harrison, Tomgally, Tommy2010, Treisijs, TuBlazin, Ulric1313, Unschool, UtherSRG, VI, Vanka5, Viajero, Vicarious, Voronov, WWGB, Waggers, Waqaspuri, Wavelength, Wayward, Wdwjbk, WendellPhillips, Whatanews4u, Wik, Winchelsea, XP, Xezbeth, Yekrats, Yidisheryid, Yusup, Zfr, Zhou Yu, Zigger, Zstroup22, Zzuuzz, ~shuri, 432 anonymous edits

Achrafieh *Source*: http://en.wikipedia.org/w/index.php?title=Achrafieh *Contributors*: A.K.Khalifeh, Abdullahslh, Alain08, Berytech, Beyrouth, Bonadea, Cedrus-Libani, Charoog10, Crzrussian, Davidcannon, DennisIsMe, Desperado lb, Elie plus, George, Geotheone101, Grm wnr, Grutness, Haddiscoe, Harryboyles, Igorliv2007, Linaduliban, Mrsimonmourtada, Murtasa, Parnellg, Pouffa, Qasamaan, Racconish, Rich Farmbrough, Rjwilmsi, Robertissimo, SamDeed, Senori, SimonP, Splash, Storkk, Supertouch, Supreme Deliciousness, Toufic1972, 83 anonymous edits

Lebanon *Source*: http://en.wikipedia.org/w/index.php?title=Lebanon *Contributors*: -- April, -Majestic-, 16@r, 1930fwc, 1execl, 1pezguy, 28421u2232nfenfcenc, 334a, 3attriss, 4destruction, 661kts, 7aresslubnan, 849fgt, 9258fahsflkh917fas, A Gooner, A Werewolf, A.h. king, A0207, A3RO, ABF, AHM, AJWittenberg, Aafour, Aaker, AaronMicael, Aaronjsussman, Abbass.sharif, Abdullah Geelah, Abdulnour, AbeCooper, Abes123, Abhijitsathe, Abidaoud, Abolabo3, Abolhood, Aboloai, Aboluay, Aboumael, Academic Challenger, Acather96, Acs4b, Addbb, Addoula, Adoniwy, Aeusoes1, Aexon79, AgadaUrbanit, Agthorr, Ahoerstemeier, Ahuskay, Ahusni, Aidanthingy, Aim Here, Aivazovsky, Akamad, Akanemoto, Akerans, Al Ameer son, Al-Andalus, Al-Zaidi, AladdinSE, Alain08, Alanderri, Alansohn, Albenali, Alberrosidus, Aldaou, Alecco23, AlefZet, Aleron235, Alex '05, AlexKiddHelicopter, Alexander Domanda, Alexandru Stanoi, Alexb102072, AlexiusHoratius, Alfonso El-Granadi, Ali@gwc.org.uk, Allstarecho, Alpha Quadrant (alt), Alste123, AM86, Amazonien, Amcleese, Amit A., AnOicheGhealai, Anahon, Aniten21, AnnieHall, Anojan123, Anonymous editor, Anonymous251, Antandrus, Anthony, Anthonyd3ca, Antogonist, AntonioBu, Antonykahwagi, Apparition11, Applebob, Aquafire information, Aquarelle, Arab Hafez, ArabianClaymore, Aranes, Aranherunar, ArchonMeld, AreaControl, Argyll Lassie, Arjun01, Arnnra, Art LaPella, Arthena, Asa Zernik, Asav, Ashmoo, Asocall, Assyria 90, Asterion, Astral, Asxasx, Athenean, Atif.t2, Atitarev, Atmleb, Atrix20, Attilios, AuburnPilot, Avala, Avenged Eightfold, Avenue X at Cicero, Averver, Avraham, Avs5221, Awesomo144, AxG, Axelschentscher, Aybray, Aym24, AzaToth, Azdro, BCtl, BEIRUTMotherofGOD, BRG, BTTNext, Babahadi, Babajobu, Babbage, Baby Luigi, Bakabaka, Baldati.com, Balsa10, Banaticus, Banzoo, Barbood, Baristarim, Baronkj39, Barryz1, Bash, Basketball110, Bassamelmortada, Bastin, Batmanand, Bazonka, Bekiroflaz, Beland, Ben Ben, Benne, Bento00, Bertilvidet, Beryte, Bestofmed, Beyrouth, Bggoldie, Bhny, Big Adamsky, Biggity, Biruitorul, BjornVDM, BlastOButter42, BlazeJag, Blessthishouse, Bletch, Blicks23, BlingBling10, Bloggingbeirut, Bludvigs, BlurTento, bmwwf91, Bob O'Bob, Bobblewik, Bobo192, Boing! said Zebedee, Bokpasa, Bolivian Unicyclist, Bongomanrae, Bongwarrior, Benjamin1, Borjeily, Borkificator, Bože pravde, Bragr, Brambleclawx, Brazzouk, Brendanconway, Brheed Zonahp84, Brian the Editor, Brianski, Brilliant Pebble, Brion VIBBER, BritishWatcher, BrokenSegue, Bronzmajom, Brucelee, Bsadowski1, Btball, Buaidh, Buchanan-Hermit, Bulaybil, Bulgab, Burboosh, Bwildasi, Bydand, Bzeek, C'est moi, C.Fred, CALR, CARPEDIEM, CBF28, CIreland, CLW, CWii, Cab88, CalJW, Calabraxthis, Calbaer, Caloox, Cam, CambridgeBayWeather, Can't sleep, clown will eat me, Canadian-Bacon, Canderson7, Canglesea, Cantus, Capecodeph, Caponer, Capricorn42, Captain Baldur, Captain panda, Captainm, CardinalDan, Carlita, Carlossuarez46, Carolehokayem, Casper2k3, Catbar, Catgut, CatherineMunro, Cbet202, Cbirdsong, Cburnett, Ccarroll, Ccmancool901, Cdarstar, Cedrus-Libani, Celkadri, CentaMary, Cftrcp, Chaldean, Charles Richardson, CharlesKiddell, CharlesMartel, Charoog10, Chart123, Chase me ladies, I'm the Cavalry, ChaseS08, Chdsandeep, Ched Davis, Cheifsguy, Chocolateboy, Chowbok, Chrabiehroy, Chris the speller, ChrisCork, Chrisjj, Chrism, Christian75, ChristianH, Christopher Lims, Chun-hian, CieloEstrellado, Cirt, Cjreeve, Ck1914, Clarince63, Clicketyclack, Closedmouth, Cluckbang, Cmf04d, Cncs wikipedia, CodeName 88, Codf1977, Cohesion, Collins.mc, Colonies Chris, Cometstyles, CommonsDelinker, Compoundinterestisboring, Conversion script, Copana2002, Coppertwig, Corpx, Courcelles, Cpl Syx, Cptnono, Crass Spektakel, Crazy Ivan, Crazyartsyguy, Crazymonkey1123, Crazyvas, Credema, Crediblerumination, Cs-wolves, Cshobe, Cst17, CuteHappyBrute, Cybercobra, Cyberzizo, Cyde, Cygnebleu, Cylon, CylonCAG, Cynops.pyrrhogaster, Cyraq, DJ Clayworth, DRosenbach, DVD R W, DVdm, DaGleamingPhoenix, Daa89563, Daanschr, Dadude3320, Dahveed323, Danger, Daniel Case, Daniel Olsen, Daniel Quinlan, Daniel5127, Daniel575, DanielCD, Danielcohn, Danny, Daqron, Darko3d, Darth Panda, Darwinek, DaveBurstein, David Betesh, David Kernow, David Koller, David Underdown, David hx, Davidcannon, Davidpdx, Dawnseeker2000, Dbachmann, De Administrando Imperio, De728631, Deadcorpse, Deanb, Debresser, Deflective, Dekimasu, Dekisugi, Delirium, Delldot, Den fjättrade ankan, Dendodge, Denghu, Denisarona, Deodar, Der Eberswalder, DerHexer, Derekmcmillan, Deville, Digitalme, Dimadick, Dina, DinajGao, Dinopup,

Gebran_Tueni *Source*: http://en.wikipedia.org/w/index.php?title=Gebran_Tueni *Contributors*: Ad vitam aeternam, Adashiel, Advisor, AnonMoos, Arab Hafez, Babajobu, BanyanTree, Bbsrock, Berasategui, Betacommand, Bobblewik, Captainm, Cedars009, Cedrus-Libani, Chzz, Delirium, Deville, DuKot, Emileassi, Eranb, Eternalsleeper, Fero45, Firevai, Fjmustak, FrancescoMazzucotelli, Freddy123, Gabolos, Geni, George, Good Olfactory, Hiram111, Jaxl, Kbdank71, Kernel Saunters, Khoikhoi, Klaksonn, Louaymk, Lugnuts, Mani1, MarkGallagher, Marudubshinki, Michaeladenner, Miss-simworld, Nurg, PBP, Petergee1, Pharos, Pissant, Prester John, Rich Farmbrough, Rjwilmsi, Robofish, SGBailey, Stefanomione, Thameen, Treybien, Vpendse, Vuvar1, Yalibnan, Ynhockey, Youssefaoun, Zalali, 67 anonymous edits

Lebanese_Forces *Source*: http://en.wikipedia.org/w/index.php?title=Lebanese_Forces *Contributors*: 4destruction, A.h. king, Abou-3akar, Aboumaroun, Alansohn, Aldis90, Alyoshka, Anon anon, Anon anon anon, Appetite4D, Ardric47, Arre, Asbestos, Assyria 90, Bernardd55, Cam, Captainm, Cedrus-Libani, Charbelkaram, Charles-LF, Chendy, Cherry blossom tree, Closedmouth, Coulonges, Danarz, Danrolo, Dark 182 88, Darko3d, Davidcannon, Deconstructhis, Degen Earthfast, Delirium, DickClarkMises, Donbosco9, Douaihy, Eatcacti, Eik Corell, Electionworld, Elie plus, Elie.Fares, Elie1234321, Elvis214, Epbr123, Equitor, Eternalsleeper, Ettrig, Faylasoof, Fedayee, Flewis, FunkMonk, Gaff, Geotheone101, Geronimo81, Glass Sword, Gotread, HanzoHattori, Heraclius, Hunter., Husond, IngerAlHaosului, Iridescent, IronyWrit, Ixfd64, Izzedine, J04n, J10452M, JacquesBittar, Jarry1250, Jdheyerman, JoeThePublic, John of Reading, Johnaelias, Johnpacklambert, Jopsach, Julekmen, JustAGal, Kablammo, Kataebi, Katimawan2005, Katous1978, Kenaani, Kravath, Kubigula, Kukini, Lanternix, Le jugle, Lebanonassist, Leszek Jańczuk, Lfeditor, LilHelpa, Linuxbeak, Lizrael, Lo2u, MER-C, MLeb, Magioladitis, Magister Mathematicae, Maniac18, Maradite, Marek69, Marioz, Mboverload, Michel10452, Mickel123, Mild Bill Hiccup, Miss-simworld, MitsosGreece, MorisSlo, Nick Number, Nv8200p, Omicronpersei8, OwenX, Patrickury, Paul August, Petronas, Philopedia, PhnomPencil, Plastikspork, Prince Cadmus II, Qasamaan, R Lee E, RJFF, Rafy, Ralhazzaa, Ranndy.Smith, Rbrwr, Rich Farmbrough, Ricomalko, Rjwilmsi, Rm uk, Robin Hood 1212, Salimzwein, Sasquatch, Schmiteye, Seraphimblade, Simplyskillz, Skarioffszky, Sl1982, Socrates2008, Soman, Spirit slash, Stormie, Sunil060902, Tewfik, The Gnome, TheDJ, Theda, Think Fast, Thuresson, Tualha, Usurp, WJBscribe, Wanabka, Werldwayd, WhisperToMe, Whiteabbey, WikiDao, Wilfried Derksen, Will314159, WisamFarouk, Wjmtruepatriot, Woodyjames29, Woohookitty, Xtcrider, Zeno Gantner, Zerolando, 739 anonymous edits

Image Sources, Licenses and Contributors

Printed by Books on Demand GmbH, Norderstedt / Germany